Redefining the Electron

Dilip D James

2024

Foreword

The study of electrons has long been at the forefront of scientific inquiry, shaping our understanding of the universe and its fundamental laws. As we delve deeper into the intricacies of quantum mechanics and particle physics, we find ourselves confronted with concepts that defy conventional logic, challenging the very fabric of our understanding. In this context, "Redefining Electrons" emerges as a beacon of innovative thought, introducing ground breaking concepts that compel us to rethink our approach to the natural world.
Dr J Cottaral

Introduction

The story of the electron is a fascinating journey through the annals of scientific discovery. From its conceptual roots in ancient Greek philosophy to its formal discovery in the late 19th century, the electron has consistently challenged our understanding of the natural world. Yet, despite the substantial body of knowledge accumulated over the past century, the electron remains an enigmatic entity, its true nature is often obscured by the complex tapestry of quantum mechanics. This book, "Redefining the Electron", aims to provide a fresh perspective on the electron, questioning established theories and offering novel insights that challenge conventional wisdom.

The need for a new interpretation of the electron is rooted in the shortcomings of current models. The wave-particle duality, while instrumental in the development of quantum mechanics, presents a paradoxical view that often defies intuition. How can something be both a particle and a wave? This conundrum has puzzled scientists and philosophers alike, leading to a variety of interpretations that seek to reconcile the seemingly contradictory behaviour of electrons. The complexity and abstract nature of these theories have prompted calls for a simpler, more coherent explanation.

"Redefining the Electron" addresses this need. Dr D D James

This book is dedicated to Dr M R Srinivasan, doyen of physicists, colleague and companion of Nobel Laureate Dr Homi Bhaba, friend to Dr Vikram Sarabhai founder of ISRO (Indian Space Research Organsiation). To Dr M R Srinivasan, my sincere thanks and appreciation for your support, advice and help

Table of Contents

Chapter 1 : Some Questions and Answers

Some questions and answers

This is a book that poses questions on fundamental concepts in physics. Questions such as why is it possible to shield an object from Electromagnetic waves but not from gravity? Or do distances really contract in special relativity and does time really dilate?

Did Max Planck really prove that energy was discrete and that it arrived in discrete, quantized packets of energy rather than as a continuous flow? Yet another question has to do with the size of waves, although waves look small to us being about one ten millionth of a meter or ten thousandths of a millimeter in size, from the point of view of an atom, they are inordinately large. What is the wavelength of a wave? Simply put the wavelength of a wave is the distance over which a wave's shape repeats itself. If one looks at the glass in the microwave door, it is possible to see that the glass is lined with a grid or mesh, the purpose of this grid is to protect one from the harmful microwave radiation in the oven, based on the size of the microwave wave length. The size of the grid is calculated so that it blocks the microwave radiation coming from the

1

oven. But what does the size of a wave have to do with the atom? It should be noted at this juncture that although modern scientists tend to dismiss the concept of the electron having any size, it is interesting to note that since the electron possesses mass it is a corollary to this that it must also possess dimensions. Using classical electrodynamics and x-ray spectroscopy, the classical radius of the electron can be calculated as being approximately 2.817×10^{-15} meters. If one uses this figure of the electron size, the results are both relevant and interesting. For instance, if Max Planck's theory of quantized units of energy is true, how does an electron that is 2.8×10^{-15} m in size (classical radius of the electron) absorb an incoming photon with a 500 nm wave-length that is 1.78×10^{8} times its size. If one tries to solve this problem mathematically, it yields an impossibly large number:

Size of electron $= 2.8 \times 10^{-15\,m}$

Size of incoming photon/electromagnetic wave $= 5 \times 10^{-7}$ m

Therefore difference in size between incoming photon (electromagnetic wave of 500 nm size) $= (5 \times 10^{-7})/(2.8 \times 10^{-15}) = 178 \times 10^{6}$ that is to say the incoming wave-length is one hundred and seventy eight million times the size of the electron.

If one considers this fact, it is an impossibly large difference in size. So the question arises, does the incoming photon or electro-magnetic wave possessing a

2

wave length of 500 nm interact with an individual electron OR does it interact with the atom in which the electron is OR is it interacting with all the atoms in the substance with which it (the 500 nm wavelength wave) is interacting? Even if the quantum mechanics assertion that the electron when in the atom is a wave-function and not a particle, is taken into account, no sense can be made of the huge disparity in size of the incoming 500 nm wavelength photon.

An atom has an approximate size of only 10^{-10} m approximately; therefore a very large discrepancy in size exists between the incoming 500 nm wavelength and the atom with a size of 10^{-10} m.

Area of 500 nm squared = $(5 \times 10^{-7})^2 \times 3.14 = 7.85 \times 10^{-13}$ m^2

Area of atom = $(10^{-10})^2 \times 3.14 = 3.14 \times 10^{-20}$ m2

Number of atoms in 5×10^{-7} m^2 = (7.85×10^{-13}) / (3.14×10^{-20}) = 25×10^6 atoms

Therefore number of atoms in the 7.85×10^{-13} m^2 area occupied by an incoming 500 nm wave is equal to twenty five million atoms approximately. This is a large number.

So the question is, which of these 25×10^6 (Twenty five million) atoms does the incoming 500 nm photon interact with? Which electron in that one out of 25 million atoms, does the incoming 500 nm wave length photon finally interact with to be absorbed and re-emitted?

3

According to quantum mechanics or rather the Standard Model which includes all the various disciplines of quantum mechanics including quantum electrodynamics (QED) and quantum field theory (QFT) the manner in which the above logistically impossible feats are achieved is explained in a **perfect** manner, by using appropriate mathematical models. Whether this statement proves to be true or not will be examined later on during the course of this book.

There remain yet more, many more questions to raise about fundamental concepts of physics and still more questions related to the electron. It must have surely come to the notice of many people, that mobile phones are capable of processing information at the rate of several Gigahertz. This is what makes it possible to watch streaming content over the net and store and watch films in high definition. Think about what the word 'processing' connotes: it describes a process wherein data is collected, evaluated, manipulated and stored. To imagine that the tiny processor in a smart phone is able to achieve all of these tasks at the rate of gigabits per second is nothing less than incredible. Yet when one considers the concept of photon emission as explained by the Standard Model it is found that the concept of photon emission by a bound electron is an indeterminate one-off affair. Here is a quote from physics forums:

"Electrons emit photons through a process called spontaneous emission. This occurs when an electron in an

4

excited state spontaneously transitions to a lower energy state, releasing a photon in the process."

This is hardly descriptive of the way in which photons are seen to be emitted, the evidence points to the fact that photons should be emitted not spontaneously but at rates in the trillions of Hertz.

There also arise, among other things in questions related to fundamentally held concepts in physics the question of the CMBR. For instance take the question of the Cosmic Microwave Background radiation (CMBR), is it plausible? The Cosmic Microwave Background Radiation is thought to be relic radiation left over from the time of the Big Bang or more accurately from a time of about 300, 000 years after the Big bang. Does it mean that the Universe in the present day is absolutely quiescent? If one stands next to the Ocean, it has a definite sound, yet the Universe we live in is supposed to be so silent that the Cosmic Background Microwave Radiation can be clearly heard. Further, modern telescopes have enabled astrophysicists to view light from Galaxies situated 13.5 billion light years distant. A difficult task, yet the CMBR that evolved at roughly the same time can be heard by just tuning into an old valve type radio. A signal arriving from such a distance should theoretically need the most sophisticated equipment we possess. The most important aspect of this book is that it not only poses questions about fundamental concepts in physics but that it also provides answers to those questions. In order to achieve such an end

5

in a meaningful way, the unusual step will be taken of first elucidating the theory that provides answers to these questions, and only later going into justifications and proofs.

The New Theory that will be illustrated in this book is known as Gestalt Aether Theory or more Simply as A New Theory of the Aether:

The word gestalt means something that is greater than the sum of its parts. Gestalt Aether theory was evolved primarily to address the issue of the propagation of light, but went on to answer several other problems. What is Gestalt Aether Theory and how can it be considered to be a Theory of Everything? Gestalt Aether Theory represents a new paradigm in the history of physics. By looking at previously collected data in a radically new way, Gestalt Aether Theory both rationalises and explains all of the different phenomena in physics under a single heading. Therefore, the propagation of light, an explanation of gravity, an explanation of magnetism, electricity and super conductivity is achieved using a single paradigm. Physical phenomena can be calculated using prevailing mathematical formula as a first approximation, making it a versatile, accurate and verifiable theory.

Nothing in science is certain, any theories that have been proposed and formulated have been founded on observation and experiment and if the observations and experiments explain the phenomenon accurately a new theory is created. However, sometimes it becomes necessary as new facts come to light, to alter or discard old theories in favour of new ones.

Yet this process is not always infallible and not always followed; sometimes a theory gets so mired down in usage that even though it seems to go against circumstantial evidence to the contrary, it is still taken to be the more acceptable solution. There are many instances of this kind in present day physics where better, proven alternatives have been passed over in favour of established customs and practices. In this book several such instances are examined: they include electric and magnetic fields, the propagation of light, the propagation of gravity, the propagation of electricity, the theory of wave-particle duality and several other such phenomena that should if reason is followed be abandoned or re-examined due to new hard or even circumstantial evidence pointing to other explanations than existing ones.

Redefining the Electron

In the period beginning with the start of the twentieth Century and continuing till the present day, the science of physics has shown a steady change in ontology; from an empirical approach to an increasingly metaphysical one. The result of such an approach has been to manage complications to a theory, or unresolved questions, with a 'leave it alone, it will solve itself approach'. This has resulted in the irreversible forward march of physics with unresolved or inadequately explained concepts lying like unexplored islands, in its wake.

Take for instance, the question of the size of photons. A photon with a wave length of 500 nm is approximately 178 million times larger (approx.) than the classical radius of the electron at 2.18×10^{-15} m and if an average atom can be taken to be about 1×10^{-10} m on diameter the incoming 500 nm wave length would be approximately 5000 times larger in size than the atom. How can this question, of the discrepancy in size between incoming photons and electrons and atoms be resolved? (This book presents a compelling solution for the discrepancy between photon size and electron size.). Max Planck, the physicist who discovered that energy was discrete and not continuous, was quite definite that any solution involving electromagnetic radiation involved discrete individual quanta of energy that we know of as photons.

Max Planck 1856 - 1947

Neils Bohr, with his meticulous research into atomic spectra during the formulation of his model of the atom, also proved exactly the same point in a practical manner with his experiments on atomic spectra. Even, today, with all the sophisticated technology available to bring to bear on this problem, it is unresolved but paradoxically, is also deemed unworthy of discussion. Does the incoming wave interact with all the atoms in a substance, and then in some strange and mysterious way, get absorbed by the correct electron, in the correct atom? Even the description of the electron as an 'electron cloud' or more accurately as a wave-function, does little or nothing to ameliorate the situation.

It is like the 18th Century metaphysical philosopher Bishop Berkeley's contention, that when there was no one in the room to perceive it, all the furniture in the room vanished and only reappeared when someone else entered the room.

George (Bishop) Berkeley 1685 - 1753

George (Bishop) Berkeley was an Anglo-Irish Anglican bishop, and metaphysical philosopher, and scientist best known for his empiricist and idealist philosophy, which holds that reality consists only of minds and their ideas.

In other words, it was Bishop Berkeley's contention that objects only had existence when they were perceived. Take as another example, of this kind of metaphysical logic, the use of Schrodinger's equation. Schrodinger's equation and the mathematics arising out of it, underpin a large part of quantum mechanics, yet there are problems. Schrodinger's equation comes up with an almost perfect solution for a single particle system, but then with each additional particle that is added to the system; an additional three spatial dimensions are required. An accurate description of Hydrogen with its single electron can be easily depicted with the 3-dimensional wave function that we are familiar with, Yet the wave function for the two electrons of the helium atom could not be interpreted as two three dimensional waves existing in ordinary three-dimensional space. Instead, the mathematics pointed to a single wave inhabiting a strange six-dimensional space. In each move across the periodic table from one element to the next, the number of electrons increased by one and an additional three dimensions were required.

Erwin Schrodinger 1887 – 1961

It is puzzling to know why such an equation, with its obvious faults continued to find favor. It is probable that the conception of the electron as a standing wave

11

within the atom was a very tempting solution to the objection raised by Larmor of the electron spiraling into the nucleus. The problem of extra dimensions that resulted from the application of Schrodinger's equation to multiple electron atoms still persisted even after the paradoxical solution of treating the Schrodinger wave-equation as an abstract mathematical function to determine the probability of finding a particular electron within the atom. When, applied to light with its almost infinite number of particles the number of extra dimensions becomes infinite. This problem was solved by Max Born stating that the wave function was not something that was real but that it existed as an abstract mathematical wave function. However, Erwin Schrodinger, experienced such great disappointment when he realized that his concept of the electron as a standing wave did not describe reality that he left the study of physics altogether and took up his old profession as a teacher of music.

The reason for addressing these issues at this juncture is because the present book seeks to present a new theory called Gestalt Aether Theory that offers coherent and cohesive resolutions to problems such as photon-to-electron size discrepancies and the propagation of light, without resorting to drastic measures such as the use of creation and annihilation operators.

Definition of the electron:

Electrons are subatomic particles that are fundamental to the structure of matter. They are one of the primary constituents of atoms, along with protons and neutrons. Electrons carry a negative electric charge, and they orbit the nucleus of an atom in specific energy levels or shells.

Although the literature and information on the electron is vast, a closer examination of the electron and its properties shows that several key factors arising out of the electrons observed properties have either been ignored or misinterpreted. For instance, both the charge of the electron and its mass are known with a great degree of accuracy. What has not been discussed is HOW the electron maintains its charge. Physicists are aware of the fact that whenever the charge on an electron is measured, either through physical experiment or through inductive reasoning from its interaction with other substances, that the electron charge is always 1.6×10^{-19} C. The question that has never been asked is, how does this happen? We are aware that the electron is always changing its energy levels and that it inevitably returns to its base energy of 1.6×10^{-19} J. Why and how does this happen? It has also been noted that the changes in the electrons energy levels are due to interactions with quantum excitations of the electromagnetic field. Could it be that the electrons interactions are superficial and therefore its core energy is

13

never involved? Therefore, it is clear that the present wisdom sees the photon as existing externally to the electron. The electron and the photon are separate entities, that only come together when they interact.

If one examines these conclusions of the Standard Model of quantum mechanics, namely that the electron is a separate entity and the photon is a separate entity. The realization begins to dawn that this explanation does not fit in with what has been observed and learned over the past half a century. It is known, from the research of Max Planck and others that energy as it has to do with light is quantized very specifically. For instance, one can look at the frequency of light travelling in a vacuum (or even through air) and state exactly what the wave-length and energy of the light is. Yet, this observation does not hold good when applied to the same light seen as a stream of photons: the concept of frequency does not exist for photons in the way that it applies to waves. The second part of the observations made about the electron, is that these interactions (photon emissions and absorptions) take place at the rate of several hundreds of trillions of times per second. This is no longer conjecture but a proven fact as demonstrated in the working of the latest, rubidium optical atomic clocks.

Electrons oscillate at the rate of hundreds of trillions of times per second. The mechanism that has been put in place in the Standard Model to explain photon emission and absorption does not explain this

14

phenomenon. The Standard Model of photon emission is that an electron in an atom that is receptive to certain energy levels interacts with an excitation of the electromagnetic field possessing that particular energy level. The electron gains energy from the interaction and moves to a higher energy level, and after some undetermined time, the electron falls back to a lower level and emits a photon equal in energy to the difference in energy levels within the atom that are involved. This explanation does not explain the rate of photon emission that takes place. Further, the very idea of the electron existing within the atom as a wave-function that changes shape resulting in the emission of a photon is not adequate. How can one wave (the wave-function of the electron) interact with another wave (the electromagnetic photon wave) to produce yet another electromagnetic wave having a specific energy, frequency and wave-length? The Standard model mathematical explanation of the interaction between the incoming electromagnetic wave (photon) and the electron wave-function is that it is explained by the 'interaction Hamiltonian'. Quantum mechanics mathematics based as it is on imaginary numbers can hardly be expected to depict reality. Yet, this is precisely what the proponents of quantum mechanics have believed to be the reality for more than three quarters of a century. The obvious question that arises out of these observations is: Does an alternative theory exist? Fortunately such a theory does exist, and it offers cogent explanations for electricity, for magnetism, for photon emission and absorption, for the propagation of light, for the formation of black holes, for the nature of Dark Matter, and finally for the long searched for explanation of gravity. The name of the theory is Gestalt Aether Theory or simply a New Aether Theory.

Gestalt Aether Theory : The photon its structure and properties

The study of light has probably occupied the human mind for almost as long as its own existence. It is therefore appropriate that prior to any discussion of the photon, to look more closely on observations that have been made of its properties. While the photon itself is of relatively recent origins, many of the properties of light had already been noted.

Unfortunately, no very cogent explanation exists for what photons are, or of how photons come into existence. Where are photons stored? If they are stored in the field what complicated mechanism exists that result in an excitation of the field of the needed energy to produce a photon of the needed frequency? How do photons exist in such a wide variety of frequencies, wave-lengths and energies, numbering in the hundreds of trillions of frequencies energies and wave-lengths? At present photons are thought to have no independent existence but are

16

treated as excitations of the magnetic and electric fields, manifesting themselves when and where they are needed. In investigating a possible structure and form for the photon, attention was naturally drawn to the work of Crick and Watson, who had undertaken a similar search on the structure of the DNA molecule. In the case of the DNA molecule, too, where purely theoretical and mathematical based investigations had met with failure, a more practical approach resulted in success. Watson & Crick had decided to discard a purely theoretical approach and to opt for a more hands on approach. A similar decision was made in the search for photon structure outlined in this book. Initially this process began with creating a list of the observed properties of the photon:

1. A photon has no mass.
2. A photon is never still it always travels at the speed of light c when it is in a vacuum.
3. A photon is electrically neutral.
4. A photon has a fixed energy that it maintains intact.
5. A photon can be emitted and absorbed by electrons.
6. Photons can be present in trillions of frequencies and wave lengths.
7. Photons with a frequency of the visible spectrum are emitted directly by electrons.
8. Radio-waves are formed by a different process and can reach lengths of 5,000,000 m or more.

   A significant thought that occurs when looking at the work of Neils Bohr and other scientists who

worked with atomic spectra, is that photons are intimately connected with electrons. The electron is a charged particle, could the possibility exist that the photon is also made up of electric charge? The idea that the photon is composed of electric pulses emitted by the electron was investigated and found to be a viable possibility.
Look at Figure 1:

+

Pulses of electrical energy emitted from electron are polarised

Figure 1.

It was realised that the initial pulses of energy emitted by the electron might be stronger than subsequent pulses of energy emitted by the electron resulting in polarisation of the emitted pulses of electric energy. The individual pulses of energy emitted by the electron are separated by a space from which all matter has been displaced forming a perfect dielectric. Figure 2.

Solenoidal field takes shape around pulses of energy forming a photon

Figure 2.

Lastly, and perhaps the most important aspect of the photon structure and formation process discussed in this section, is the question of the size of the photons formed by electrons. To begin with all photons, whether virtual photons or real photons, possess the identical structure as is Shown below in Figure 3 below.

Figure 3

The size of the above depicted photon has a diameter of approx. 10^{-16} m and a length of approx. 10^{-6} m. The following points should be considered:

a) This structure of the photon means that the photon possesses no mass. It is a massless 'particle'.

b) This model of the photon is electrically neutral. It has the structure of an electric dipole that is electrically neutral and physically stable.

c) This model of the photon has a fixed energy that is maintained intact from the time it is emitted to the time it is absorbed.
d) This model of the photon has a stable configuration that cannot be easily disrupted.
e) This model of the photon explains how optical photons are produced directly by the electron. Higher energy photons such as gamma rays and lower energy photons such as radio-waves have a similar but different genesis. Gamma rays are formed during the destruction of the nucleus while radio waves are formed by a different process that shall be explained in a subsequent chapter.

Therefore this model of the photon fulfills all of the requirements and physical properties that a photon is supposed to possess. Further:

1) This size of the photon means that it can easily be formed and emitted by an electron in a time interval of 10^{-18} s.
2) The fact that an individual electron can deal on a one on one basis with incoming photons, means that it is possible for the electron to oscillate at the rate of several hundreds of Terahertz per second and to absorb and emit photons at that rate.

3) This extremely high rate of oscillation means that it is possible to explain the rectilinear nature of light, a property that accounts for, among other things, the casting of shadows. This is explained by the fact that for as long as the electron is being irradiated by photons from a given direction and source, it will continue to emit photons at the rate of 10^{14} Hz or more, all of which are identical to each other in terms of frequency, wave-length and energy. These photons will all be emitted in keeping with the classical laws of reflection where $\angle$ i $= \angle$ r. (angle of incidence equals angle of reflection). This results in rays or lines of hundreds of trillions of connected photons being formed per second all of which are identical to each other, creating a ray of monochromatic light.

4) This structure of the photon as a stable electric dipole also explains how light propagates according to the inverse square law, since it enables the photon to link up with other photons. (This process will be explained in more detail at a later point.).

5) Most importantly, it explains the manner in which photons are linked to electrons in a comprehensible and vivid manner instead of

depending upon the explanation that photons are the result of quantum fluctuations in the electromagnetic field.

6) According to Gestalt Aether Theory electric and magnetic fields do not exist neither does an electromagnetic field; only the virtual photon field exists that serves both electric and magnetic functions. An electric field consists of a field of polarized virtual photons and a magnetic field consists of an energized field of virtual photons, wherein energy is flowing.

7) The physical space that exists between pulses of electrical energy emitted by the electron during the formation of a photon, have all molecules and atoms displaced. Hence representing a pure dielectric between the pulses of energy. Allowing the photon to maintain its energy intact for very long periods of time.

In conclusion to this Section, it should be stated that Gestalt Aether Theory has evolved a very satisfactory model for the structure and formation of the photon. According to the Gestalt Aether Theory, photons have their Genesis inside electrons; they are not something apart and external as theorized by Quantum mechanics. Gestalt Aether Theory makes the photon an integral part of the electron, light (photons) is therefore an integral property of electrons and not a result of the interaction of electrons with excitations of the electromagnetic field as is currently held by quantum mechanics. In order to make this concept more easily understood, a mathematical description of the process is included.

Before going on to the mathematical description of the forming of a photon, the main postulate of the Gestalt Aether Theory should be stated:

"The electron always tries to maintain its base energy of 1.6 x 10^{-19} J intact, if it gains excess energy through absorption of a photon it immediately sheds that excess energy through emission of a photon. Therefore, the net energy of the electron remains unchanged. "

What does this mean? It means that an electron regulates its energy by absorbing or emitting photons that suit its energy requirements and that maintains its base charge of 1.6 x 10^{-19} C. Looking at an electron, this can be expressed mathematically as:

Electron Energy Invariance Equation:

$$E_0 + \sum_{i=1}^{\infty} (E_{\text{photon}} - E_{\text{emission}}) = E_0$$

This equation indicates that an electron's energy remains constant over time taking into account emission and absorption events.

In order to incorporate the charge invariance of the electron, differential equations can be used to model the

24

rate of change of energy and the rate of absorption and emission of photons by the electron.

Rate of Change of Energy:

Let $E(t)$ represent the energy of the electron at time t.

The rate of change of energy of the electron can be described by the equation:

$$\frac{\partial E}{\partial t} = K_{abs} \times f^2 h(t) \times E(t)$$

$$(2)$$

$$K_{em} = K_{abs}$$

Where:

K_{abs} is the rate constant for absorption of photons.

K_{em} is the rate constant for emission of photons.

f is frequency

h is planck's constant

t is time

$f^2 h(t)$ is the intensity of incident photons over time t.

2. *Rate of Absorption and Emission:*

The rate of absorption of photons by the electron is proportional to the frequency of incident photons over time multiplied by Planck's constant, which can be expressed as:

$$K_{abs} = \alpha \cdot f^2 h(t)$$

$$(3)$$

Similarly, the rate of emission of photons by the electron is proportional to the frequency of photon absorption, which can be expressed as:

$$k_{\text{em}} = \beta \cdot E(t)$$

$$(4)$$

3. *Equilibrium Condition:*

The system reaches equilibrium state when the rate of absorption equals the rate of emission:

$$k_{\text{abs}} = k_{\text{em}}$$

$$(5)$$

$$\alpha \cdot f^2 h(t) = \beta \cdot E(t)$$

$$(6)$$

This describes the continuous absorption and emission processes taking place within the atom involving electrons and photons. The rate of change of energy of the electron is governed by the balance between the rate of absorption and the rate of emission of photons. The equilibrium condition ensures that the system always reaches a steady state where the rate of absorption equals the rate of emission.

By solving these differential equations and considering appropriate initial conditions and parameters, it is possible to analyze the dynamics of energy exchange between electrons and photons in various scenarios. This model provides a quantitative understanding of the continuous interaction between electrons and photons, which is essential for studying phenomena such as light absorption, emission, and scattering.

Focussing solely now on the electric field and its interactions as described by this new theory. The mathematical representation of a photon is as follows:

26

1. Emission of Pulses of Electric Energy:

Let $E(t)$ represent the strength of the electric pulses emitted by the electron as a function of time (t).

The initial pulse of energy can be represented as $E_0(t)$, and subsequent pulses can be represented as

$E_i(t)$, where ($i > 0$).

2. Polarization of Energy Pulses:

The polarization factor for each pulse can be denoted as P_i, where P_0 represents the polarization factor for the initial pulse, and P_i represents the polarization factor for subsequent pulses $i > 0$.

3. Formation of a dipole Electric Field:

The polarization of the energy pulses results in a dipolar electric field forming around them. This electric field can be represented by a vector field $\vec{E}$.

The magnitude and direction of the electric field depend on the polarization factors Pi of the energy pulses.

4. Stable Configuration

The stable, massless, electrically neutral configuration is achieved when the dipole electric field balances the external electric field generated by the pulses of energy.

Using mathematical notation:

Initial pulse of electric field: $E_0(t)$
Subsequent pulses of electric field: $E_i(t)$ for $i > 0$
Polarization factor: P_i for $I = 0, 1, 2, \ldots$.

Vector field representing solenoidal electric field: $\vec{E}$
Magnitude and direction of $\vec{E}$ depend on P_i for each pulse.
Stable configuration achieved when the dipole electric field achieves a stable configuration, where it is essentially in a neutral state.

Photon emission and absorption process:

The physical process of photon emission and absorption by the electron will now be examined in a little more detail. This new theory of the Aether, explains sub-atomic processes involving the emission and absorption of photons within the atom as follows. The theories adopted by the Standard Model with its heavy dependence on wave-particle duality and its depiction of electrons moving within the atom as an electron cloud is dismissed and instead the electron is treated as a solid particle orbiting the nucleus. As had been illustrated earlier it is thought that the electron oscillates at a very high rate and emits and absorbs photons at the rate of several hundreds of trillions of photons per second. In doing this the electron follows the laws of classical physics as applied to recoil processes where $\Theta_{incidence} = \Theta_{recoil}$. If one cares to look at everyday life, it is noticed that absorption and emission take place over very long periods, in fact

28

absorption and emission processes continue for as long as excitation of the electron within the atom takes place. Take for example excitation by artificial light or by sun-light, electrons are excited and are oscillating at a rate of hundreds of trillions per second, for very long periods of time. How is this possible? It is held by this new theory that since the electron and the proton, possess equal and opposite charge that when they make contact, they neutralise each other. This makes it possible for an excited electron to physically recoil off the nucleus. Examining a single interaction of an electron with an incoming photon, this is what takes place. A radiating source emits a photon, note that the size of the photon with a diameter of 10^{-16} m makes it possible for a specific electron in a specific atom to easily absorb and emit specific photons, the emitted photon is absorbed by an electron with the proper energy level within the atom. The absorption of the photon imparts momentum to the electron propelling it in the direction of the nucleus. Since the size of a proton (nucleus) is approximately 2000 times the size of the electron, the nucleus appears to the electron to be a perfectly flat, perfectly smooth surface. As the electron approaches the nucleus, there is an appreciable increase in its speed, instead of smashing into the nucleus, when the electron and nucleus touch, the equal and opposite charges possessed by the electron and proton are temporarily neutralised. It should be remembered that the atom is neutral because overall, the charge of electrons to protons is equal. Any extra interaction that takes place can therefore be treated as an individual interaction between electron and proton.

As a practical illustration of this process, think of an electron ejected by the cathode and travelling towards the anode, the electron does not get repelled by the anode at the last minute. A look at static electric charges serves as an even better illustration when positive and negative charges meet they neutralise each other and are no longer attracted. In the same way, the negative charge of the incoming electron and the positive charge of the nucleus which are equal but opposite are momentarily neutralised, what is left is the energy and momentum that has been imparted to the electron by the incoming photon.

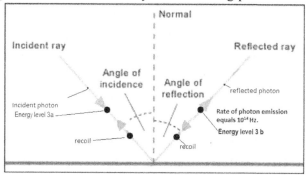

Absorption and emission of photons by electron, notice the recoil off the nucleus in keeping with the classical laws of reflection.

The electron therefore, rebounds or recoils off the nucleus, following the laws of classical recoil and

30

reflection where $\Theta_{incidence} = \Theta_{recoil}$. When the electron reaches the energy level at which it had absorbed the photon but at a location exactly opposite to its original location, it emits a photon of the same value it had absorbed. In order to cope with the forces of recoil resulting from the emission of a photon it retraces its path and ends up in its original position at n = 3, where it absorbs another photon and the whole process repeats. This process takes place at the rate of hundreds of trillions of repetitions of absorptions and emissions every second and continues for as long as the excitation from that particular source continues. If the energy of the emitted radiation from the source continues and the direction remains unchanged, rays of connected photons are created that travel in straight lines. This process explains the rectilinear nature of light.

The following is a mathematical explanation illustrating the process of photon emission and absorption:

Suppose that the atom under consideration is a hydrogen atom and that the incoming photon has a wavelength of 700 nm.

Given

$$\lambda = 700. \ nm = 700 \times 10^{-9} \, m$$
$$c = 3 \times 10^{8} \, m/s$$
$$h = 6.62607015 \times 10^{-34} \, m^{2} \, kg/s$$

First calculate the frequency:

$$f = \frac{c}{\lambda} = f = \frac{c}{700 \times 10^{-9}}, = 4.28 \times 10^{14}$$

$$(7)$$

Calculate the energy E of the photon:

$$E = hf = (6.62607015 \times 10^{-34}) \times (4.2857 \times 10^{14}) = 2.8385 \times 10^{-19} J$$

Convert energy to electron volts eV:

$$1eV = 1.602176634 \times 10^{-19} J \therefore 2.8385 \times 10^{-19} J \times 1.602176634 \times 10^{-19} = 1.772 \, eV$$

$$(9)$$

Solving for n energy level of the atom:

$$1.772eV = \frac{13.6}{n^2}$$
$$n^2 = \frac{13.6}{1.772} = n^2 = 7.6709$$
$$\therefore n = \sqrt{7.6709} = 2.769$$

Since n must be an integer, the nearest integer value is n = 3.

Thus, the electron would be absorbed into the third energy level orbit of the atom when a photon with a wavelength of 700 nm is absorbed. The following is the sequence of events:

Given that the absorbed photon = λ_a and emitted photon = λ_e

1. Electron Absorbs Photon λ_a (700 nm) and Gains Momentum

2. Energy of absorbed photon: $= E_a = \dfrac{hc}{\lambda_a}$

$$(11)$$

To calculate the momentum gained by the electron we can use the formula:

$$E = \dfrac{p_1^2}{2m_e}$$

$$(12)$$

Where m_e is the mass of the electron

3. $p_1^2 = 2.85 \times 10^{-19} \times 2 \times 9.10938356 \times 10^{-31}$

$$(13)$$

$$2.85 \times 10^{-19} = \dfrac{p_1^2}{2 \times 9.10938356 \times 10^{-31}}$$

$$p_1 = \sqrt{5.1709 \times 10^{-48}}$$

$$p_1 = 7.19 \times 10^{-24}\, kgm/s$$

4. The momentum gained by the nucleus p_2 is equal in magnitude but opposite in direction to p_1, so $p_2 = -7.19 \times 10^{-24}$ kg m/s. Since the nucleus is thousands of times the mass of the electron, the recoil has a negligible effect on the nucleus and a greater effect on the electron.

5. Electron and Nucleus Collision, Neutralization of Charges:

6. Charges of electron and nucleus neutralize each other: $1.6 \times 10^{-19} + 1.6 \times 10^{-19} = 0$

7. Excess momentum and energy imparted by absorbed photon remain:

8. $E_{remaining} = E_a = 1.722 \ eV$ causing it to recoil or reflect from the surface of nucleus:

9. Electron Recoils from Nucleus According to Classical Laws: Angle of incidence equal to angle of reflection: $\Theta_i = \Theta_r$

10. Electron Reaches Energy Level n = 3 and Emits Photon λ_e (700 nm):

Energy of emitted photon: $= \quad E_e = \dfrac{hc}{\lambda_e}$

(14)

The sequence repeats, with the electron experiencing gain in momentum from emission of photon at new position in $n = 3$ and undergoing recoil taking it back to its original position at $n=3$ and the whole sequence repeats at the rate of hundreds of trillions of repetitions per second.

This process of photon absorption and emission by the bound electron may be described on a step by step basis as follows:-

Electron in Stable State: An electron occupies a specific energy level within an atom, maintained by its energy, mass, momentum, etc.

Photon Absorption: The electron is irradiated by an external source, absorbing incoming photons of an acceptable energy level at the rate of hundreds of trillions of photons per second.

One at a Time: The electron deals with the incoming photons singly one at a time.

Energy Absorption: When the electron absorbs an incoming photon, it results in a net gain in energy.

Recoil Towards Nucleus: To conserve energy and momentum, the electron is propelled towards the nucleus due to the recoil from the energy gain.

Neutralization at Nucleus: Upon reaching the nucleus, the electron's negative charge is temporarily neutralized by the proton's positive charge, achieving a neutral state.

Rebound Mechanism: Despite the neutralization, the electron still retains the extra energy imparted by the absorbed photon. It rebounds off the nucleus according to the laws of classical physics, where the angle of incidence equals the angle of reflection.

Energy Emission: Upon reaching the higher energy level, the electron emits the extra energy gained by the absorption of the photon, returning to its original energy level of 1.6×10^{-19} J.

Recoil and Return Path: The emission process causes the electron to recoil, following the path it took to its original position.

Repetition: The electron continues to absorb incoming photons, repeating the cycle of propulsion towards the nucleus, neutralization, rebound, emission, and recoil, all according to the classical laws of reflection , all according to the classical laws of reflection where:

$\angle i = \angle r$.

Incorporating these new concepts into the framework of Gestalt Aether Theory emphasizes the conservation of the electron's original energy level during the emission and absorption process. This approach offers a classical perspective that seeks to integrate concepts of energy absorption, emission, and momentum conservation in a coherent framework. At the same time it also explains in a clear, easily understood manner, how electrons can be emitted and absorbed at the phenomenal rate of hundreds of trillions of photons every second, which is what is needed to explain this phenomenon in the modern context. The laconic quantum mechanics one-off description of photon emission and absorption is no longer good enough.

This description with its classical connotations provides an intuitive understanding that complements the quantum mechanical framework, offering a broader perspective on the behaviour of electrons within an atom. This is the new ontology of the electron within the atom.

Chapter 2 : Properties of the Electron and Virtual Particles

The electron is a fundamental particle, one of the building blocks of matter. Its discovery in the late 19th century revolutionized our understanding of atomic structure and led to the development of modern physics. This section explores the fundamental characteristics of the electron, including its charge, mass, spin, and behavior in various physical contexts.

Charge of the electron

The electron carries a negative electric charge, which is a fundamental property intrinsic to its nature. The magnitude of this charge is approximately -1.602×10^{-19} coulombs, denoted as (e). This quantized charge is one of the key characteristics distinguishing electrons from other subatomic particles. The negative charge of the electron balances the positive charge of protons in atoms, ensuring overall electrical neutrality.

The precise measurement of the electron's charge was achieved through Robert Millikan's oil-drop experiment in 1909. By observing the behavior of charged oil droplets in an electric field, Millikan determined the fundamental unit of electric charge, providing crucial evidence for the existence of discrete charged particles. This experiment will be described in a detail later in this Chapter

Mass of the Electron

The electron has a very small mass compared to other subatomic particles, particularly protons and neutrons. The rest mass of an electron is approximately 9.1 x 10^{-31} kilograms. This mass is about 1/1836 that of a proton, making electrons the lightest stable subatomic particles known. Despite their small mass, electrons play a crucial role in the structure and behavior of atoms. The determination of the electron's mass was a direct consequence of J. J. Thomson's experiments with cathode rays. By measuring the charge-to-mass ratio of the electron and combining this with Millikan's charge measurement, scientists could calculate the electron's mass accurately.

Spin and angular momentum of the electron

Electrons possess an intrinsic form of angular momentum known as spin. Unlike classical angular momentum, spin is a quantum property with no direct classical analog. Electrons have a spin quantum number of $\pm \frac{1}{2}$, indicating two possible spin states. These states are often referred to as "spin-up" and "spin-down." The concept of spin is integral to the quantum mechanical description of electrons. It contributes to the electron's magnetic moment and influences how electrons interact with magnetic fields and other particles. The Pauli exclusion principle, is a fundamental tenet of quantum mechanics, states that no two electrons in an atom can have the same set of quantum numbers, including spin. This principle underlies the structure of electron shells and subshells in atoms.

Wave-particle duality

One of the most profound discoveries in modern physics is the wave-particle duality of electrons. Electrons exhibit both particle-like and wave-like properties, depending on the experimental context. This duality was first suggested by Louis de Broglie in 1924 and later confirmed by experiments such as the Davisson-Germer experiment, which demonstrated electron diffraction. The wave nature of electrons is described by quantum mechanics, particularly by the Schrödinger equation, which provides a mathematical framework for understanding electron behavior in atoms. Electrons occupy quantized energy levels and orbitals, described by wave functions, which determine the probability distributions of finding electrons in particular regions around the nucleus.

Electrons in magnetic and electric fields

Electrons are charged particles and are thus influenced by electric and magnetic fields. In an electric field, electrons experience a force proportional to their charge, causing them to accelerate towards the positive electrode. In a magnetic field, moving electrons experience a Lorentz force perpendicular to both their velocity and the magnetic field direction, resulting in circular or helical trajectories.

The behavior of electrons in fields is exploited in numerous technological applications. Cathode ray tubes (CRTs), once common in television and computer monitors, rely on the deflection of electron beams by electric and magnetic fields to create images on a screen. Electron microscopes use magnetic lenses to focus electron beams and achieve high-resolution imaging of tiny structures.

The Electron and quantum mechanics

The early 20th century saw the emergence of quantum mechanics, a new framework for understanding the behavior of particles at atomic and subatomic scales. Classical mechanics, which accurately described macroscopic phenomena, failed to explain the behavior of electrons and other tiny particles. Key experiments, such as the photoelectric effect and the discrete spectral lines of hydrogen, indicated the need for a new theoretical approach.

Multiple Dimensions in Quantum Mechanics

In 1926, Austrian physicist Erwin Schrödinger formulated the Schrödinger equation, a foundational equation in quantum mechanics. This equation describes how the quantum state of a physical system changes over time. Schrödinger's approach was based on the wave nature of particles, as proposed by de Broglie, and it provided a way to calculate the probability distribution of an electron's position in an atom. The Schrodinger equation which described the electron as a standing wave within the atom rather than as a particle, worked well when applied to the hydrogen atom with its single electron, but failed in describing atoms with multiple electrons, the addition of each new electron required three additional dimensions. Therefore the helium atom required 3 new additional new dimensions and the lithium atom needed 9 dimensions. This was obviously a major impediment to the implementation of the Schrodinger wave-function as it was a description that did not reflect reality.

The interpretation of the wave function and its implications for reality posed significant philosophical challenges.

The Copenhagen interpretation

The Copenhagen Interpretation of quantum mechanics was developed primarily by Niels Bohr and Werner Heisenberg. They suggested that the wave function represents a probability amplitude, and physical quantities are inherently probabilistic until measured. This interpretation resolved the issue of multidimensional wave functions by treating them as mathematical tools for predicting measurement outcomes rather than literal descriptions of electron positions.

Various alternative interpretations of quantum mechanics have been proposed to address the conceptual challenges posed by the Schrödinger equation:

The Many-Worlds Interpretation of quantum mechanics was proposed by Hugh Everett in 1957, this interpretation suggests that all possible outcomes of a quantum measurement actually occur in branching parallel universes. This view treats the wave-function as a real entity that encompasses all possible states.

David Bohm proposed a Pilot-Wave Theory which came to be known as Bohmian mechanics. According to this interpretation, developed by David Bohm in the 1950's, pilot wave theory posits that particles have definite positions guided by a "pilot wave" described by the wave-function. This view preserves determinism and provides a clear picture of particle trajectories.

Decoherence, is the latest model adopted by quantum mechanics. This modern approach explains the apparent collapse of the wave-function as a result of the interaction between a quantum system and its environment. Decoherence theory provides a mechanism for the emergence of classical behavior from quantum systems without requiring wave function collapse.

The development of the Schrödinger equation marked a pivotal moment in the history of physics, providing a powerful tool for understanding the behavior of electrons and other quantum particles. The equation's extension to multiple dimensions introduced new interpretational challenges, which have been addressed through various interpretations of quantum mechanics. These interpretations continue to shape our understanding of the fundamental nature of reality.

The Electron in Atomic Structure

Understanding the role of the electron in atomic structure has been a central quest in physics and chemistry. This chapter traces the historical development of atomic models, highlighting key discoveries and theoretical advancements that have shaped our current understanding of the atom.

Early Atomic Models : Dalton's Atomic Theory

John Dalton

John Dalton

John Dalton was born on September 5th, 1766, in the village of Eaglesfield, Cumberland, England and died at the age of 78 on July 27, 1844, in Manchester, England. Dalton was an English meteorologist and chemist, but his most important contribution was as a pioneer in the development of modern atomic theory.

In the early 19th century, John Dalton proposed the first modern atomic theory. Dalton's model suggested that atoms are indivisible particles, each element consisting of identical atoms with specific weights. While Dalton's theory laid the groundwork for modern chemistry, it did not address the internal structure of atoms. Although Dalton might have made some errors in the details of the manner and ratio in which elements combined, his overall contribution to an eventual understanding of the atom was immeasurable.

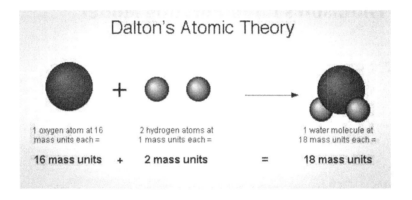

Dalton's Atomic Theory

1 oxygen atom at 16 mass units each =

16 mass units   +

2 hydrogen atoms at 1 mass units each =

2 mass units

1 water molecule at 18 mass units each =

=   **18 mass units**

Daltons Atomic Theory

The first part of his theory states that all matter is made of atoms, which are indivisible. The second part of the theory says all atoms of a given element are identical in mass and properties. The third part says compounds are combinations of two or more different types of atoms.

45

More surprises were in store however, it was found that the atom, thought to be indivisible for thousands of years, was in fact made up of smaller constituent parts.

J J Thomson

J J Thomson was born on December 18, 1856, in Cheetham Hill, near Manchester, England he passed away on August 30, 1940, at Cambridge, Cambridgeshire. J J Thomson's helped revolutionize the knowledge of atomic structure by his discovery of the electron (1897). He received the Nobel Prize for Physics in 1906 and was knighted in 1908.

Thomson's Plum Pudding Model

J J Thomson

J.J. Thomson's discovery of the electron in 1897 led to the first model of atomic structure incorporating subatomic particles. Thomson proposed the "plum pudding" model, where the atom was envisioned as a sphere of positive charge with negatively charged electrons embedded within it, similar to plums in a pudding. This model suggested that

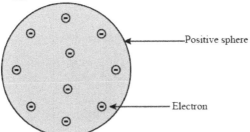

Thomson Plum Pudding model of the atom

electrons were evenly distributed throughout a positively charged matrix. Although Thomson did not follow up on his theories he did speculate that electrons might orbit the positive charge along specific orbits.

Rutherford's Nuclear Model

Born on a farm in New Zealand in 1871, the fourth of 12 children, Ernest Rutherford completed a degree at the University of New Zealand and began teaching unruly schoolboys. He was released from this task by a scholarship to Cambridge University, where he became J. J. Thomson's first graduate student at the Cavendish Laboratory. At the Cavendish Laboratory he began experimenting with the transmission of radio waves.

Ernest Rutherford

Rutherford later went on to join Thomson's ongoing investigation into the conduction of electricity through gases, and then turned to the field of radioactivity just opened up by Henri Becquerel and Pierre and Marie Curie. He was famous for his discovery of the transmutation of elements, which was conducted with his fellow researcher Frederick Soddy. Rutherford and Soddy concluded that: "The idea of the chemical atom in certain cases spontaneously breaking up with the evolution of energy is not of itself contrary to anything that is known of the properties of atoms."

A consummate experimentalist, Rutherford was responsible for a remarkable series of discoveries in the fields of radioactivity and nuclear physics. He discovered alpha and beta rays, set forth the laws of radioactive decay, and identified alpha particles as helium nuclei.

Most importantly, he postulated the nuclear structure of the atom: experiments done in Rutherford's laboratory showed that when alpha particles are fired into gas atoms, a few are violently deflected, which implies a dense, positively charged central region containing most of the atomic mass. This model of the atomic structure, came to be known as the planetary model of the atom, because of its resemblance to the solar system.

The Gold Foil Experiment

In 1909, Ernest Rutherford and his colleagues conducted the famous gold foil experiment, which involved firing alpha particles at a thin sheet of gold foil. Most alpha particles passed through the foil with little deflection, but some were deflected at large angles, and a few even bounced back. If Thomson's Plum Pudding model of the atom were true, the electrons should have passed right through without encountering any impediment. This unexpected result led Rutherford to propose a new model of the atom.

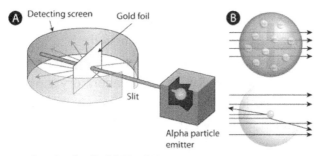

Rutherfords Gold Foil Experiment

     Rutherford concluded that the atom consists of a tiny, dense, positively charged nucleus surrounded by electrons. The nucleus contains most of the atom's mass, while the electrons occupy the surrounding space. This nuclear model replaced Thomson's plum pudding model and introduced the concept of a central nucleus. This model of the atom introduced by Rutherford is often called the planetary model of the atom since it resembles the solar system.

The Bohr Model of the atom

     Niels Bohr was born in Copenhagen, Denmark, in 1885. He was the son of Christian Bohr, who was also professor of physiology at the University of Copenhagen. Bohr's mother was Ellen Adler Bohr and was Jewish with a rich family of bankers and politicians. Bohr was a mathematician, physicist, and football player in the national team. In 1903 Bohr began studies of philosophy and mathematics at the Copenhagen University. In 1905 the Royal Danish Academy of Science organized a competition. Niels Bohr participated with some experiments on surface tension. With the use of his father's laboratory to conduct experiments Bohr managed to win the prize, which helped in his focussing his energy on the study of physics. He received his doctorate in 1911.

Bohr joined Trinity College in Cambridge in 1913, where he studied under the tutelage of J.J. Thomson. Later he went to the University of Manchester where he worked under the direction of Rutherford, after which he returned to Denmark.

Neils Bohr

Spectral Lines and Quantized Orbits

In 1913, Niels Bohr proposed a new model of the atom to explain the discrete spectral lines observed in atomic emission and absorption spectra. In both his theory and his experimentation Bohr was heavily inspired by the discovery of the Balmer series of spectral lines. The Balmer series of spectral lines was first discovered by

Johann Jakob Balmer, a Swiss mathematician who discovered a simple arithmetic formula relating the wavelengths of lines in the hydrogen spectrum in 1885. This proved to be the start of intense activity in precise wavelength measurements of all known elements and the search for general principles and was also the primary influence in Bohr's investigation into the atom. In 1888 Robert Rydberg a Swedish physicist and mathematician generalised the Balmer series through a formula that came to be known as the Rydberg formula.

Bohr's model combined Rutherford's nuclear structure with the idea of quantized energy levels. He postulated that electrons orbit the nucleus in fixed energy levels or shells without radiating energy. Electrons could transition between these levels by absorbing or emitting specific quanta of energy, corresponding to the observed spectral lines.

Electron transitions for the Hydrogen atom

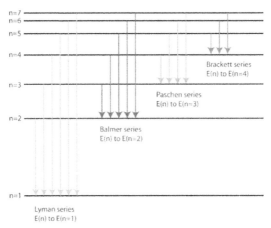

Electron transition creating spectral lines

Postulates of the Bohr Model

The Bohr model of the atom made several postulates, some of which were not sustainable.

1. Quantized Orbits: Electrons orbit the nucleus in fixed, quantized orbits with specific energies. These orbits are stable and do not emit radiation.

2. Energy Transitions: Electrons can transition between energy levels by absorbing or emitting photons of specific energies. The energy difference between levels determines the frequency of the emitted or absorbed radiation.

3. Angular Momentum Quantization: The angular momentum of an electron in orbit is quantized and given by $L = n\hbar$, where n is a positive integer (the principal quantum number) and $\hbar$ is the reduced Planck constant.

53

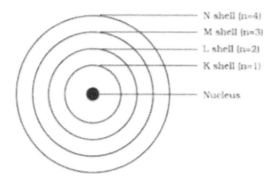

N shell (n=4)
M shell (n=3)
L shell (n=2)
K shell (n=1)

Nucleus

Bohr atom model

Successes and Limitations

The Bohr model successfully explained the spectral lines of hydrogen and provided a foundation for understanding atomic structure. However, it had limitations. It could not accurately predict the spectra of atoms with more than one electron and failed to account for the finer details of spectral lines (such as the Zeeman effect).

Impact on Atomic Theory and Limitations of the Bohr model of the atom.

The Bohr model marked a significant advancement in atomic theory. It introduced the concept of quantized energy levels, paving the way for the development of quantum mechanics. Bohr's work demonstrated the need for a more comprehensive theory to explain atomic structure and electron behaviour.

The introduction and eventual rejection of the Bohr model of the atom could also be thought to mark the end of classical physics. The concept of wave-particle duality was introduced at the 1927 Copenhagen Conference. According to the Copenhagen Interpretation, atomic and subatomic particles sometimes act like particles and sometimes act like waves. This is called "wave-particle duality." An electron, for example, when detected, is in its localized particle form. But between detected positions, an electron is in its wave-like form. The introduction of wave-particle duality by a General consensus of eminent physicists during the Copenhagen Solvay Conference of 1927, sealed the fate of classical physics and represented a clear demarcation between the old and the new physics.

The Quantum Mechanical Model

Although this book makes some harsh criticisms of quantum mechanics, it should be kept in mind that quantum mechanics has enabled some truly amazing advancements that almost eclipse the achievements of classical physics. The unleashing of the power in the atom being not the least of these achievements. The curie family consisting of Pierre and Marie Curie and their daughter Irene and son in law Frederick Joliot Curie had played a huge, often unsung part in this discovery, ultimately giving their lives to the advancement of science. While Pierre

Curie died at age 46 from an accident, Marie, Irène and Frédéric died from diseases caused by their exposure to radiation during the course of their scientific experiments. Marie Curie, working with her husband, Pierre Curie, discovered the elements polonium and radium in 1898. In 1903 they won the Nobel Prize for Physics for discovering radioactivity. In 1911 she won the Nobel Prize for Chemistry for isolating pure radium, making her the only person to win a Nobel prize in two disciplines. It was these discoveries that enabled Ernest Rutherford and Frederic Soddy to determine that the elements could undergo transmutation, changing from one element to another. They also discovered the existence of isotopes of various elements. The concept of isotopes had been previously unknown. Rutherford's most important discovery, namely the structure of the atom, later called the planetary model of the atom, would not have been possible without the presence of a source of radium to use in his gold foil experiments. It was the gold foil experiment that yielded the basic structure of the atom as being composed of a massive positively charged nucleus surrounded by mostly empty space in which were found the negatively charged electrons.

The Copenhagen Interpretation of quantum mechanics

Quantum mechanics has done a wonderful job of explaining the atom and its structure. However, it should be remembered that the part of quantum mechanics that deserves such recognition was based almost completely, on hands on empirical research and experimentation. For the first twenty-seven years, after its initial introduction in 1900 by Max Planck, quantum mechanics and classical physics co-existed side by side in harmony, both schools of thought being given equal importance and being used with equal effect. However, post the formal introduction of wave-particle duality in 1927, the quantum mechanics stand hardened, forcing the old accepted concepts of classical physics out of everyday use. The fifth Solvay physics conference held from October 24 to 29, 1927, and Chaired by Dutch physicist Hendrik Lorentz, was originally devoted to discussions on "electrons and photons" but was eventually dominated by disputes about the ideas behind quantum mechanics. The 1927 Solvay conference on physics, resulted in one of the first major interpretations of quantum mechanics and is called the Copenhagen Interpretation. During the course of this conference the transition was made from, what is often referred to as the weak interpretation of quantum mechanics to the strong interpretation of quantum mechanics. These two interpretations, the strong and the weak interpretation of quantum mechanics, have to do with the nature of the Universe. The weak interpretation of quantum mechanics was so named because it doesn't interpret anything new, it placed a lot of dependence on mathematical calculations and if any doubts arose, physicists were told to do the calculations. In other words, don't try to interpret the results, just validate them. By contrast the strong Copenhagen interpretation, held that not only does the wave-function describe the behavior of the electron but more importantly the wave-function before the collapse of the wave-function was thought to be something real that had a physical existence. According to the strong interpretation of quantum mechanics, the wave-function was something real. Therefore, while the weak interpretation of quantum mechanics treats the wave-function as an abstract tool that was not real, the strong interpretation of quantum mechanics stated that not only was the wave function real but that it had a physical existence. Therefore, light literally exists as a wave function before it is detected and collapses into a regular state.

This mind set is still very much prevalent in the present day. In his you tube video, entitled "The Multiverse is real. Just not in the way you think it is." Dr. Sean Carroll, one of the pre-eminent physicists of the day, supports the multiverse theory in a very definitive way. The multi-verse theory states that, light as it travels from point A to point B travels as a mathematical wave-function, when it is detected at point B, the wave function collapses and the light again becomes real. Concurrently with the collapse of the wave function multiple parallel

Universes are born. In the same video Dr Michio Kaku, a particle physicist of note, states that the Universe has 11 dimensions.

In this book a detailed investigation will be carried out into the reason for the introduction of wave-particle duality and more importantly whether such a concept as wave-particle duality is justified. This is because it was the introduction of wave-particle duality into quantum mechanics that led to the concepts such as the wave-function, super position, quantum entanglement and other purely quantum mechanics concepts.

Joseph Larmor (1857-1942)

Joseph Larmor, an Irish physicist, made significant contributions to the understanding of electromagnetic theory and the behaviour of charged particles. In 1897, Larmor published a paper addressing the issue of how accelerated charged particles, such as electrons, would radiate energy according to classical electromagnetism. The Problem that Joseph Larmor brought to the notice of scientists was that according to classical electrodynamics, an accelerated charged particle emits electromagnetic radiation, losing energy in the process. For an electron orbiting a nucleus, this would mean a continuous loss of energy, causing the electron to spiral into the nucleus rapidly. When Larmor performed the actual calculations he came to the shocking conclusion that an electron orbiting the nucleus should radiate away all of its energy in ten pico seconds (10^{-11} s).

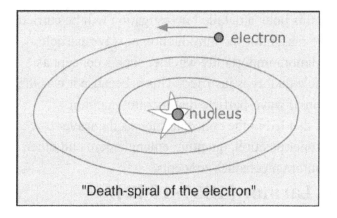

"Death-spiral of the electron"

Why don't electrons spiral into the nucleus

This was a fundamental issue as it implied that atoms, and therefore matter, would be inherently unstable. In fact a purely logical conclusion would be that neither matter nor anything else, including the earth, the solar system and the Universe should exist. The question therefore, became one, not of why the world did not exist, but of why it did. One of the solutions that was suggested for the manifestation of the world and everything else, was wave particle duality. Therefore the introduction of wave-particle duality was acknowledged as much for its philosophical explanation of how the Universe could exist as for its explanation of why electrons did not spiral into the nucleus. It is an accepted fact of life that without the adoption of wave-particle duality, the central question of how we could continue to exist would not have been answered, or at least as will be found during the course of this book, would not have been immediately answered. Because of objections like these, Bohr's model of the atom was ultimately rejected after having enjoyed more than a decade of supremacy. The reason that the Bohr model was eventually rejected was that although Bohr had postulated that electrons in certain orbits do not radiate, he could never explain in logical terms why this was so. He also could not adapt his model to explain multi electron atoms.

A short History of the events leading up to the introduction of wave-particle duality

It was a well acknowledged fact of classical physics, one that was experimentally verified, that any accelerating charge (including electrons in circular orbits) should emit electromagnetic radiation. This radiation would cause the electron to lose energy and spiral into the nucleus, in a period of time that was calculated at 10 pico seconds (10^{-11} s) or ten trillionths of a second (100,000,000,000 s). Bohr circumvented this issue by postulating that electrons in certain stable orbits do not radiate energy. These orbits were associated with quantized angular momentum. This postulate, while effective, was ad hoc and lacked a rigorous theoretical foundation.

Bohr proposed that the angular momentum of an electron in orbit is quantized and given by ($L = n\hbar$) (where (n) is an integer and ($\hbar$) is the reduced Planck constant). This quantization was an attempt to impose stability, but it was not derived from first principles. Bohr suggested that electrons emit or absorb energy when they transition between orbits. However, this idea faced scepticism because it didn't explain the mechanism behind these transitions.

Arnold Johannes Wilhelm Sommerfeld was a famous theoretical physicist known for his atomic and quantum physics contributions. He was born on 5 December 1868 in Königsberg, Prussia. Sommerfeld extended Bohr's model to include elliptical orbits, which provided a better explanation for fine spectral lines. Sommerfeld, also incorporated relativistic corrections, improving the accuracy of predictions for spectral lines, especially for elements other than hydrogen.

The major breakthrough that addressed the limitations of Bohr's model came with the development of quantum mechanics in the mid-1920s. Werner Heisenberg, at one time a close companion and assistant to Neils Bohr, introduced a new type of matrix mechanics, using non-commutative algebra to describe quantum states. This formulation avoided the need for visualizing electron orbits and instead focused on observable quantities like energy levels. This was a huge breakthrough as it considerably simplified many of the problems.

But it was Erwin Schrodinger who provided the breakthrough that was needed. Schrödinger proposed wave mechanics, describing electrons as wave functions (Ψ) where the electron was depicted as a standing wave rather than as particles in orbits. The wave function provided a probabilistic interpretation of the electron's position and momentum.

The time-independent Schrödinger equation ($H\Psi = E\Psi i$) describes the energy states of an electron in an atom. The solutions to this equation correspond to quantized energy levels, naturally explaining why electrons don't radiate energy in these states.

Both Albert Einstein and Arnold Sommerfeld were initially sceptical of Bohr's model but later contributed to the quantum mechanical understanding that supported it. Pauli's exclusion principle (1925) explained the electron configurations in atoms, reinforcing the stability of the electron structure proposed by quantum mechanics.

Among others Max Born, Pascual Jordan, and Heisenberg, further developed the matrix formulation, providing a solid mathematical framework for quantum mechanics.

In 1905, Albert Einstein published a paper explaining the photoelectric effect, in which he proposed that light consists of discrete packets of energy called photons. The energy (E) of a photon is related to its frequency (f) by the equation ($E = hf$), where (h) is

Planck's constant. He also proposed that photons have momentum, given by the equation:

$p = E/c = hf/c$ where (p) is the momentum of the photon, (E) is the energy, and (c) is the speed of light. This can be rewritten using the wavelength λ of the light:

$p = h/\lambda$

Prince Louis De Broglie

Prince Louis De Broglie

Prince Louis-Victor-Pierre-Raymont, 7th duc de Broglie, generally known as Louis de Broglie, is best known for his research on quantum theory and for his discovery of the wave nature of electrons. Louis de Broglie was born on August 15, 1892 in Dieppe, France, he died on March 19, 1987 at Louveciennes in France. De Broglie was a French physicist best known for his research on quantum theory and for predicting the wave nature of electrons. He was awarded the 1929 Nobel Prize for Physics "for his discovery of the wave nature of electrons." The wave-like behaviour of particles discovered by de Broglie, was used by Erwin Schrodinger in his formulation of wave mechanics.

Inspired by Einstein's equation for photon momentum, de Broglie proposed that if light, traditionally understood as a wave, could exhibit particle-like properties (photons), then perhaps particles could exhibit wave-like properties. De Broglie hypothesized that particles such as electrons have an associated wavelength given by: $\lambda = h/p$

De Broglie was initially hesitant to present his ideas, which were revolutionary and counterintuitive at the time. He faced scepticism and uncertainty about how his hypothesis would be received by the scientific community. Einstein recognized the significance of de Broglie's hypothesis and understood its potential to advance quantum theory. He saw de Broglie's work as a natural extension of the wave-particle duality of light, which he himself had helped establish through his explanation of the photoelectric effect. To encourage de Broglie, Einstein travelled to Paris in 1924 to meet him personally. During this meeting, Einstein urged de Broglie to present his ideas formally, emphasizing the importance of his hypothesis and its implications for physics. Einstein also communicated with other prominent physicists to garner support for de Broglie's work. His endorsement was instrumental in ensuring that de Broglie's ideas received the attention they deserved.

Encouraged by Einstein, de Broglie included his hypothesis in his doctoral thesis, titled "Recherches sur la théorie des quanta" ("Research on the Theory of Quanta"), which he submitted to the University of Paris in 1924. De Broglie successfully defended his thesis in front of a panel that included notable physicists. His thesis was initially met with scepticism, but the endorsement from Einstein helped persuade the scientific community of its validity. Paul Langevin, a prominent French physicist and one of the examiners of de Broglie's thesis, played a crucial role in recognizing the importance of de Broglie's work. Langevin's support further legitimized de Broglie's hypothesis.

Albert Einstein's proactive efforts to encourage Louis de Broglie were pivotal in bringing de Broglie's revolutionary ideas to the forefront of physics. Einstein's support not only helped de Broglie overcome his reluctance but also ensured that his hypothesis on wave-particle duality received the recognition it deserved, ultimately leading to significant advancements in quantum mechanics. It is indeed an established fact that Einstein's equation for photon momentum offered a conceptual loophole that Louis de Broglie utilized to propose wave-particle duality. Let's delve into how this unfolded:

In 1905, Albert Einstein published a paper explaining the photoelectric effect, in which he proposed that light consists of discrete packets of energy called photons.

The energy E of a photon is related to its frequency f by the equation $E = hf$, where h is Planck's constant.

Albert Einstein

Einstein also proposed that photons have momentum, given by the equation:

$p = E/c = hf/c$ where p is the momentum of the photon, E is the energy, and c is the speed of light. This can be rewritten using the wavelength (λ) of the light:

$p = h/\lambda$

De Broglie hypothesized that particles such as electrons have an associated wavelength given by: $\lambda = h/p$ where λ is the wavelength, h is Planck's constant, and p is the momentum of the particle.

The Conceptual Loophole, left by Einstein's equation for photon momentum suggested that waves (light) could have momentum, a traditionally particle-like property. This blurred the distinction between waves and particles, suggesting a dual nature. De Broglie extended this concept by proposing that particles, which traditionally were understood to have only particle-like properties, could also have wave-like properties. De Broglie's hypothesis was revolutionary because it generalized the wave-particle duality beyond photons to all matter. This was a logical extension of Einstein's earlier work. In other words De Broglie posited that any moving particle, not just photons, has a wave associated with it. The wavelength of this wave is inversely proportional to the particle's momentum.

De Broglie's hypothesis bridged a gap that had existed in quantum theory, providing a new way to understand the dual nature of matter. Hitherto, Schrodinger's wave function was a purely theoretical concept, with the introduction of De Broglie's Wave-particle duality, a physical basis existed in support of the theory. The hypothesis was confirmed experimentally by the Davisson-Germer experiment in 1927, which showed that electrons produce diffraction patterns, a behaviour characteristic of waves. De Broglie's work laid the foundation for the development of wave mechanics by Erwin Schrödinger, who formulated the Schrödinger equation describing how the wave function of a quantum system evolves. The concept of wave-particle duality became a cornerstone of quantum mechanics, fundamentally altering our understanding of the microscopic world.

Is wave-particle duality justified?

But how true was wave-particle duality, and is it justified? In order to determine the answer to this question, an investigation into the adoption of wave-particle duality that initially began with De Broglie's discovery of the matter-wave duality concept, will be undertaken. But before we do that, we have to examine whether wave-particle duality achieved what it set out to do, namely explain why the electron, a charged particle, did not radiate away all of its energy and fall into the nucleus. Bohr had postulated that electron in certain orbits did not radiate energy.

In an exactly similar manner, wave-particle duality as conceptualised in Schrodinger's wave equation, stated that electrons do not radiate while in stable, quantized energy states for the following reasons: Stationary states are stable and do not involve the kind of acceleration that causes radiation in classical physics. Quantized energy levels ensure that electrons can only absorb or emit energy in discrete amounts during transitions between these levels, not continuously. Therefore, wave particle duality allowed or enabled the use of wave functions that in turn served to ensure that stationary states are stable, with a probability distribution that does not change over time, leading to no net change in charge distribution and hence no radiation.

Quantum Electrodynamics (QED) which was largely the inspiration of Richard Feynman further explains that in these stable states, interactions with the electromagnetic field do not lead to the net emission of real photons. These principles were thought to collectively resolve the problem of electrons spiralling into the nucleus and provide a robust explanation for the stability of atoms. But is this actually true, does the wave function provide a comprehensive explanation for the fact that electrons in the atom are stable?

A closer explanation of the quantum mechanic's explanation for the stability of the atom, show that there is considerable scope for doubt. For instance, the reason is that Schrodinger's wave equation requires 3N spatial dimensions for every extra particle that is added. Therefore, when considering atoms with multiple electrons it means having to account for multiple spatial dimensions. For instance in a lithium atom containing 3 electrons, 9 new spatial dimensions would be needed together with new spin dimensions. In considering these extra spatial dimensions one has to take into consideration that throughout the history of physics no-one has been able to physically identify or describe more than the three physical and one temporal dimension of which we are aware. These are, length, breadth, depth and time. Taken together these dimensions can describe any spatial situation, including the question of where and when. Therefore to freely talk about multiple dimensions that can neither be described

70

nor shown to exist is a problem. It follows that in atoms with multiple electrons, the wave function becomes significantly more complex. Each electron's wave function must be considered in the context of the other electrons, leading to a many-body problem.

The total wave function for a system with (N) electrons is a function of all spatial and spin coordinates of the electrons. For (N) electrons, this wave function depends on (3N) spatial dimensions and (2N) spin dimensions. Surely this does not make sense in any logical way whatsoever? This is how quantum mechanics equivocates:

The Schrödinger equation for such systems is solved using approximations and computational methods due to the complexity. Techniques like the Hartree-Fock theory and Density Functional Theory (DFT) are used to manage this complexity. These methods supposedly handle the high-dimensional space in which the wave functions are defined. While the number of dimensions may be large, the solutions provide accurate predictions for real physical properties.

Mathematical models versus physical reality

Quantum mechanics uses mathematical abstractions to model physical systems. These abstractions, like high-dimensional spaces or complex wave functions, are tools to understand and predict physical behaviour. They are not literal descriptions of physical reality but are effective in representing and solving physical problems. Can one really put much faith in such a system?

The dimensions and solutions used in these models correspond to measurable quantities and observable phenomena. The physical predictions made by these models have been extensively validated through experiments. Although these statements are made with a considerable measure of confidence, the reality is that they are indefensible. Even though quantum mechanics insists that complete transparency is used and a clear statement of the limitations and liabilities of the methods used is stated in any explanation. It still does not make the system any more ethical, it remains a highly questionable and debatable use of mathematics and logic. The final word on the use of such tactics goes something like this.

In classical electrodynamics, a radiating electron is one that is accelerating. In a stationary state, the electron's wave function is stable and does not involve the kind of acceleration that leads to radiation. The electron in a stationary state is not undergoing the type of acceleration that classical physics predicts would cause it to emit radiation. Instead, it is in a non-radiating quantum state. This statement is supported by Schrodinger's wave equation, but since Schrodinger's equation utilises 3 N spatial dimensions is it any more acceptable than Bohr's postulate that electrons in certain orbits do not radiate?

Examining Schrodinger's wave function in a little more detail and taking it step by step: The solutions to the Schrödinger equation for an electron in an atom are called stationary states or eigenstates. These states have well-defined energy levels. When an electron is in a stationary state, its probability distribution (described by the square of its wave function, (λ^2) does not change over time. Hence, the electron is in a stable configuration. (i.e., does not radiate.). These stationary states correspond to discrete energy levels. Unlike in classical mechanics, where an electron could have a continuous range of energies, quantum mechanics restricts the electron to specific energy levels. The energy difference between these levels means that the electron can only absorb or emit energy in discrete amounts (quanta), corresponding to transitions between these levels.

In classical electrodynamics, a radiating electron is one that is accelerating. In a stationary state, the electron's wave function is stable and does not involve the kind of acceleration that leads to radiation. The electron in a stationary state is not undergoing the type of acceleration that classical physics predicts would cause it to emit radiation. Instead, it is in a non-radiating quantum state.

However, conceptually, the very fact of a wave-function undergoing discrete changes in energy levels, is difficult to accept, even given the fact that it is standing waves that are involved. Again, the idea of a wave-function being able to change shape and in this way to

73

emit an electron with a definite energy seems to lead outside the limits of accepted physics and into a paranormal type of situation. The fact that this photon emission is triggered by the absorption of another wave further heightens the gap between acceptable scientific principles and quantum mechanics theory.

The Schrödinger Equation

The quantum mechanical model of the atom is primarily based on the Schrödinger equation, formulated by Erwin Schrödinger in 1926. The time-independent Schrödinger equation for an electron in a potential $V(\mathbf{r})$ is given by:

$$[-\frac{\hbar^2}{2m}\nabla^2\psi + V(\mathbf{r})\psi = E\psi]$$

where $\hbar$ is the reduced Planck's constant, m is the mass of the electron, ∇^2 is the Laplace operator, Ψ is the wave function, $V(\mathbf{r})$ is the potential energy, and E is the energy eigenvalue.

The wave function Ψ describes the probability distribution of an electron's position. The square of the wave function's magnitude, Ψ^2, gives the probability density of finding the electron at a particular point in space. The solutions to the Schrödinger equation for the hydrogen atom yield a set of quantum numbers that describe the properties of atomic orbitals:

The Principal Quantum Number (*n*): Indicates the energy level and size of the orbital. It can take positive integer values (n = 1, 2, 3, ...).

1. The Angular Momentum Quantum Number l : Indicates the shape of the orbital. It can take integer values from 0 to n-1 . Each value of l corresponds to a specific type of orbital (s, p, d, f).

2. The Magnetic Quantum Number (m_l): Indicates the orientation of the orbital in space. It can take integer values from -l to +l .

3. The Spin Quantum Number (m_s): Describes the intrinsic spin of the electron. It can have two possible values: +1/2 (spin-up) or – 1/2 (spin-down).

Atomic orbitals are regions of space around the nucleus where electrons are likely to be found. The quantum numbers determine the size, shape, and orientation of these orbitals:

Spherical Symmetry (s-orbitals): l = 0 , spherical in shape, with n determining the size. Dumbbell Shape (p-orbitals): l = 1 , two lobes along specified axes, (m_l) determines orientation. Complex Shapes (d and f orbitals): Higher (l) values lead to more complex shapes, including d and f orbitals.

According to the Standard Model and in the opinion of present day ethos in physics, the quantum mechanical model has been crucial in explaining atomic properties and behaviours such as chemical bonding, molecular structure, and spectroscopic transitions. It has also gone a long way towards explaining wave-function collapse and decoherence.

Decoherence has been used to understand the possibility of the collapse of the wave function in quantum mechanics. Decoherence does not generate an actual wave-function collapse. It only provides a framework for apparent wave-function collapse, as the quantum nature of the system "leaks" into the environment.

(note: However, taking into consideration that the wave function has recourse to multiple spatial dimensions and the creation of multiple Universes every time the wave function collapses, how much credence can be placed in a theory that is supposed to explain the physical side of nature?)

This section covers the transition from the Bohr model to the quantum mechanical model of the atom, focusing on key concepts and the Schrödinger equation.
Electron Configuration and the Periodic Table

At first sight it would appear that in spite of its inconsistencies, quantum mechanics had brought a lot to the structure of the atom and indeed much is owed to the quantum mechanics interpretation of atomic structure. That having been said, it should be remembered that change is the basis of progress. Electron configuration refers to the distribution of electrons in an atom's orbitals. It follows from the principles of quantum mechanics and determines the chemical properties of elements. This section explores how electron configuration is organized in the periodic table and its significance in understanding atomic behaviour. The following are some of the rules governing atomic structure that were introduced by quantum mechanics.

The Afbau principle states that electrons fill atomic orbitals of lowest energy first before occupying higher energy orbitals. This principle guides the sequence in which electrons are added to atoms as atomic number increases.

The Pauli exclusion principle states that no two electrons in an atom can have the same set of quantum numbers. Specifically, within an orbital, electrons must have opposite spins (one spin-up and one spin-down).

Hund's rule states that electrons occupy orbitals of the same energy (degenerate orbitals) singly before pairing up. This rule maximizes the total spin of electrons in a subshell, leading to more stable configurations.

Electron configuration is typically expressed using the following notation:

1) The Principal Quantum Number (n): Indicates the main energy level (shell) occupied by the electron.

2) The Angular Momentum Quantum Number (l): Specifies the shape of the orbital (s, p, d, f).

3) The Magnetic Quantum Number (m_l): Determines the orientation of the orbital in space.

4) The Spin Quantum Number (m_s): Describes the spin of the electron (+1/2 or -1/2).

Periodic Table

Dmitri Mendeleev was a Russian chemist who lived from 1834 to 1907. He is considered to be the most important contributor to the development of the periodic table. His version of the periodic table organized elements into rows according to their atomic mass and into columns based on chemical and physical properties. The periodic table organizes elements by increasing atomic number and groups them based on similar chemical properties. Electron configurations reflect periodic trends such as atomic size, ionization energy, and electron affinity.

Representative Examples

1. **Hydrogen (H)**: $1s^1$- Single electron in the 1s orbital.
2. **Helium (He)**: $1s^2$- Two electrons filling the 1s orbital.
3. **Carbon (C)**: $1s^2 2s^2 2p^2$ electrons filling the 1s, 2s, and 2p orbitals.
4. **Oxygen (O)**: $1s^2 2s^2 2p^4$- Eight electrons filling the 1s, 2s, and 2p orbitals.

Periodic Trends

A closer examination of the periodic tables shows that, Atomic Size increases down a group due to added electron shells and decreases across a period due to increased nuclear charge. Similarly, the Ionization Energy, decreases down a group and increases across a period due to effective nuclear charge. The Ionisation energy may be described as the minimum energy that an electron in a gaseous atom or ion has to absorb to come out of the influence of the nucleus, the atom in which the electron was original found then becomes an ion . Further the Electron Affinity which is the ability of a neutral atom to gain an electron, generally becomes more negative (more favorable) down a group and varies across a period based on atomic properties.

Applications and Modern Insights

Electron configuration is fundamental to predicting chemical reactivity, bonding, and molecular structure. Advances in computational chemistry and spectroscopy continue to refine our understanding of atomic behaviour and periodic trends.

The above explanation of electron configurations in atoms, is fairly definitive. However, it should be remembered that much of the input governing electron configuration was from the study of Chemistry based on empirical experiments. The other half of this problem based on the De Broglie concept of wave-particle duality and the Schrodinger wave-function will be examined more closely in the next Chapter, as will the logical and epistemological reasons in continuing to utilise these concepts.

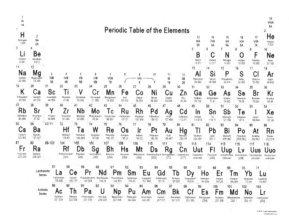

The periodic table

Chapter 3 :
Virtual particles, Gestalt Aether Theory and the Aether

Virtual particles

This section gives an in-depth look into 'Virtual' particles and the role they play in present day physics. This is something that has never before been attempted in a manner that would make the subject comprehensible to a broad audience.

Before the advent of quantum mechanics, classical physics didn't entertain the concept of virtual particles. The understanding of particles and fields was deterministic, and the notion of particles existing in a transient, virtual state was absent. However, with the development of quantum mechanics a new ethos emerged in which all possibilities were examined. Quantum mechanics could be thought to have its beginnings in 1900 when Max Planck introduced and convincingly provided experimental and theoretical proof of energy as being discrete or quantized as opposed to smooth and continuous as had hitherto been believed.

Neils Bohr

Neils Bohr's model of the atom was first introduced in 1913, and raised our understanding of the structure of the atom to new levels. It was a model of the atom that enjoyed wide-spread acceptance and acclaim. However, Bohr's atom model had flaws that were exploited by critics, for instance it was pointed out that there was no explanation as to why electrons in accelerated orbits around the nucleus did not radiate away their energy and fall into the nucleus. Another flaw in the theory was Bohr's inability to extend his model of the atom to atoms containing multiple electrons. So while the Bohr model worked well for the Hydrogen atom with its single electron and proton it failed to explain more complicated atoms.

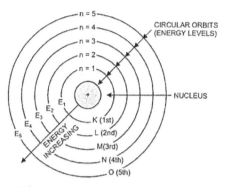

Orbit-Like Representation of Various Energy Levels

In Bohr's model of the atom the energy of an electron remains constant as long as it stays in the same orbit. This led to the idea that each orbit is associated with a definite energy, i.e., with a definite whole number of quanta of energy. The orbits, therefore, are also known as energy levels or energy shells. Bohr's Atomic Model gave numbers 1, 2, 3, 4, etc., (starting from the nucleus) to these energy levels, as shown in the above figure. These are now termed as principal quantum numbers. The various energy levels are also designated by letters K, L, M, N, etc. The farther the energy level from the nucleus, the greater is the energy associated with it.

In the 1940's Richard Feynman working closely with the Japanese Physicist Sinitro Tomonaga, developed Quantum Electrodynamics, which mathematically formalized the behavior of electromagnetic interactions. Feynman's diagrams introduced the concept of virtual particles as intermediate states in particle interactions.

84

Feynman's diagrams provided a visual representation of particle interactions, where virtual particles appear as internal lines representing transient states that mediate interactions between real particles. Physical confirmation of these ideas was yet to be found.

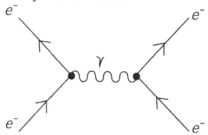

A Feynman diagram the squiggly line between the two particles represents a virtual photon.

The confirmation of the existence of the 'virtual' interactions predicted by Feynman and Tomonga was found by Willis Lamb and his partner Retherford. Willis Lamb was born on July 12, 1913 in Los Angeles, California, and died on May 15, 2008 in Tucson, Arizona after a long and respected career as a physicist. Willis Lamb and co-recipient, Polykarp Kusch, were awarded the 1955 Nobel Prize for Physics for experimental work that spurred refinements in the quantum theories of electromagnetic phenomena.

In 1947, Willis Eugene Lamb an American physicist and his partner Retherford, were working on investigating the energy levels of atoms when irradiated with microwave frequencies. By plotting the critical magnetic field strength for a variety of microwave frequencies, Lamb and Retherford could determine the energy difference between the two states in the absence of a magnetic field. Contrary to expectation, the difference was not zero. It was eventually determined that The Lamb shift is caused by interactions between the virtual photons created through vacuum energy fluctuations and the electron as it moves around the hydrogen nucleus in each of these two orbitals.

Before the discovery by Lamb and Retherford of the Lamb shift, there was theoretical speculation about the existence of intermediate states in particle interactions, primarily driven by the development of Quantum Electrodynamics, However, these speculations were more abstract and mathematical rather than focused on concrete empirical evidence. The Lamb shift provided the first clear experimental indication that virtual particles had observable consequences, solidifying their role in modern physics. Even more important was that the Lamb Shift was a reproducible experiment that could be repeatedly verified. In summary, while the concept of virtual particles was not explicitly speculated upon before the advent of Quantum Electrodynamics (QED), the theoretical framework laid by early quantum theorists paved the way for their acceptance. The Lamb shift discovery was crucial in transitioning virtual particles from theoretical constructs to accepted components of quantum field interactions.

Willis Lamb and Robert Retherford's Work in 1947 on the energy levels of hydrogen atoms provided crucial empirical evidence for the existence of virtual particles, solidifying their place in quantum electrodynamics (QED). Here's a detailed look at how their work led to the conclusion that virtual interactions were responsible for the observed phenomena.

The experiment that came to be known as the Lamb shift was motivated by discrepancies between the predicted and observed energy levels in hydrogen atoms. The Dirac equation predicted that the 2s and 2p orbitals of hydrogen should have the same energy due to their identical principal quantum number (n=2) and different angular momentum quantum numbers.

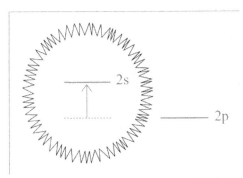

Current explanation of the Lamb shift – Schematic illustration of the Lamb shift of the hydrogenic 2s 1/2 state relative to the 2p 1/2 state. Intuitive understanding of the shift as due to random jostling of the electron in the 2s orbital by zero-point fluctuations in the vacuum field.

Lamb and Retherford used microwave spectroscopy to measure the energy difference between the $2s_{1/2}$ and $2p_{1/2}$ states of the hydrogen atom. They applied a microwave field to induce transitions between these states and measured the frequency of the absorbed or emitted radiation. They found a small but significant shift in the energy levels: the $2s_{1/2}$ state was slightly higher in energy than the $2p_{1/2}$ state. This energy difference, about 1057 MHz, came to be known as the Lamb shift. According to the prevailing wisdom the two energy levels should have been identical, the fact that there was a discrepancy in their energies indicated additional factors at work.

Quantum Electrodynamics, which had been developed jointly by Richard Feynman, Julian Schwinger, and Sin-Itiro Tomonaga, provided a framework for understanding interactions between charged particles and the electromagnetic field. QED introduced the concept of virtual particles—transient fluctuations in the field that mediate forces between particles. In the context of the Lamb shift, the energy levels of the electron in the hydrogen atom are influenced by interactions with virtual photons. These virtual photons are emitted and reabsorbed by the electron, causing slight shifts in its energy levels. The 2s and 2p states are affected differently by these virtual interactions because of their different spatial distributions and interactions with the vacuum fluctuations of the electromagnetic field. The 2s state, being closer to the nucleus and experiencing more intense vacuum fluctuations, is shifted more than the 2p state. The lamb shift, therefore, not only demonstrated that electrons interacted with virtual particles but that they did so at all spatial levels of the atom.

According to QED, the vacuum is not empty but filled with transient virtual particles that constantly appear and disappear. The key to understanding the Lamb shift lies in the concept of vacuum fluctuations. These fluctuations affect the energy levels of particles, such as electrons in an atom. The presence of virtual particles introduces radiative corrections to the energy levels predicted by the Dirac equation. These corrections account for the interactions of the electron with the sea of virtual particles in the vacuum. The Lamb shift provided strong empirical support for QED, validating the theory's predictions about the behavior of electrons and their interactions with the electromagnetic field. Willis Lamb was awarded the Nobel Prize in Physics in 1955 for his discoveries related to the Lamb shift, highlighting the significance of his work in advancing our understanding of quantum mechanics and electromagnetic interactions.

In summary, Lamb and Retherford's experiment revealed an unexpected energy shift in hydrogen atoms, which couldn't be explained by existing theories. The interpretation of this shift in the context of QED, involving virtual particle interactions and vacuum fluctuations, provided a comprehensive explanation. This work not only confirmed the existence of virtual particles but also cemented QED as a robust theory in quantum physics.

Conceptual problems arising out of the Lamb Shift:

So far during the course of this Chapter we have learnt about the Lamb shift and the discovery of virtual particles interacting with electrons and the existence of a sea of virtual particles that exists throughout the Universe. What we now need to do is to see whether the conclusions that were drawn from the proven existence of virtual particles, were indeed the correct conclusions or if the discovery of virtual particles can be viewed from other angles and viewpoints.

Even an acolyte to physics might be aware that the explanation put forward by quantum mechanics to explain how electrons interact with virtual particles is inadequate to the point that they are meaningless. What exactly do such constant exchange of virtual particles between electrons and vacuum fluctuations achieve? Is it just an amusing observation as quantum mechanics seems to think, with no deeper connotations? Or are these virtual interactions detected by the Lamb shift merely a pointer to the existence of such interactions, and to the existence of a vacuum filled with an underlying background energy that exists in space throughout the entire universe?

If one considers this subject at depth, it soon becomes apparent, that what is taking place when an electron emits and absorbs a 'virtual' photon is anything but random and anything but trivial. It can be seen that these emissions and absorptions of virtual particles by the electron take place across the entire spatial area of the atom, from being close to the nucleus as in the 2s orbit to occurring near the valence shell of the atom as evidenced by the 2p orbital interactions. One possibility that is very much in evidence is that the electron is self-stabilising its orbit by means of constantly emitting and absorbing virtual photons. This circumstance, leads to two conclusions, the first is that the electron cannot emit and absorb random energy fluctuations; in order to maintain its equilibrium the electron would have to emit and absorb very definite energies that would cause it to maintain a stable equilibrium. Therefore, here at long last is an explanation to the question raised by the Irish physicist Larmor as to why the electron which is in an accelerated state as it orbits the nucleus, does not radiate away all of its energy and fall into the nucleus. By constantly emitting and absorbing virtual photons that are undetectable, the electron is self-regulating its energy and remaining in a stable orbit around the nucleus.

The electron self-regulates its energy

The staggering implications of this conclusion are that the whole notion of wave particle duality is false.

Wave-particle duality is no longer needed to explain the question of how the electron maintains its stability within the atom: which is the primary reason that wave-particle duality had been introduced in the first place. However, before such an overwhelming conclusion is reached, it would be as well to examine in a little more detail, both the energies and the magnitudes involved in the lamb shift. It is possible that, the photons that are observed in the Lamb shift and that are thought to be virtual, are photons of normal energies but emitted over such short time intervals of time (10^{-15} s) that the Conservation laws, which are fundamental, overlook these interactions since they occur over such a short time frame that they are considered virtual processes where these laws appear not to be violated. The physical resultant of such extremely short duration emissions, are the small shifts in position detected by the Lamb shift as changes in microwave frequency. This can be explained as follows: The interaction with virtual photons perturbs the electron's energy levels. This perturbation causes a slight shift in the energy levels, known as the Lamb shift. These energy shifts are not a result of direct photon emissions observed in typical transitions but are secondary consequences of the electron's continuous interaction with the virtual photon field. The actual physical verification of the existence of virtual interactions and virtual particles, justifies, the concept of a universal virtual photon aether comprised of very low energy virtual photons each with an energy of only 10^{-51} J that were formed at the time of the Big Bang and permeate every part of the Universe, including all matter and the vacuum. This universal virtual photon field forms the background fabric of the Universe.

Introduction of virtual particle into physics of the nucleus.

Virtual particles are temporary particles that exist for a very short time. They are not directly observable but play a crucial role in particle interactions. Think of them as fleeting fluctuations that appear and disappear quickly. In the nucleus, virtual particles are responsible for mediating the fundamental forces. For example, the strong nuclear force, which holds protons and neutrons together in the nucleus, is mediated by virtual particles called gluons.

Strong nuclear force:

Protons in the nucleus repel each other due to their positive charge. However, the strong nuclear force, mediated by virtual gluons, overcomes this repulsion and holds the nucleus together. This force is much stronger than the electromagnetic force but acts over very short distances.

The weak nuclear force

The weak nuclear force, responsible for radioactive decay, is also mediated by virtual particles, specifically W and Z bosons. This force changes one type of particle into another, like turning a neutron into a proton.

Later it became apparent that even protons and neutrons cannot be considered to be fundamental but that they are made up of quarks. Quarks are held together through the exchange of pions. This interplay of forces and the exchange of virtual particles ensure that atomic nuclei are stable structures, capable of existing despite the repulsive and attractive forces at play.

Here, it is interesting to note that the introduction of virtual particles into nuclear theory took place only in 1964 with the introduction by Murray Gellman and George Zweig of quark particles, this was more than a decade after the Lamb shift had been discovered. It is noted with some amazement, that while the fact that forces in the nucleus could be mediated and held together by the exchange of virtual particles was accepted, that the same idea was ignored in explaining how the atom itself stayed intact.

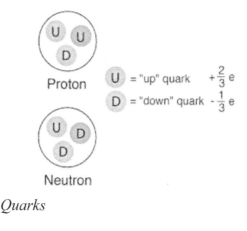

Proton

U = "up" quark $+\frac{2}{3}$ e

D = "down" quark $-\frac{1}{3}$ e

Neutron

Quarks

A quark is one of the fundamental particles in physics. They join to form hadrons, such as protons and neutrons, which are components of the nuclei of atoms. The study of quarks and the interactions between them through the strong force is called particle physics. The antiparticle of a quark is the antiquark. Quarks and antiquarks are the only two fundamental particles that interact through all four fundamental forces of physics: gravitation, electromagnetism, and the strong and weak interactions.

Why was the possibility that virtual interactions were responsible for the atoms stability ignored?

There were several reasons that the Lamb Shift and the discovery of virtual particles did not receive wider publicity and took so long to reach the public forum. Chief of these was the fact that security during the Second World War was at unprecedented levels. The atomic bomb was dropped on Hiroshima in 1946, Willis Lamb discovered the Lamb shift in 1947, only a year after the Atomic bomb had been detonated; it is no wonder that no news of his discovery leaked out: secrecy and security were at peak levels. Therefore, the discovery of virtual particles was restricted to a small group of allied scientists on a strictly need to know basis. Also the discovery of the Lamb Shift was regarded at the time as little more than a curiosity, the virtual interactions were thought to be random and unorganized, occurring whenever electrons in the atom interacted with random fluctuations of the vacuum. Therefore, the idea that the exchange of virtual particles by electrons within the atom could actually displace wave-particle duality, never even came to mind.

The Lamb Shift was initially regarded as a curiosity due to its unexpected nature and the limited understanding of virtual particles at the time. Virtual particles were thought to be involved in random fluctuations rather than organized exchanges, which constrained the exploration of alternative theories. The concept of wave-particle duality remained dominant, as it was a well-established explanation fitting the existing quantum mechanical framework, without considering the potential of virtual interactions to displace it.

This historical perspective helps understand why the implications of virtual particles were not fully explored in the context of the Lamb Shift and why the established concepts of wave-particle duality continued to prevail.

If one considers quantum mechanics it is possible to see that it is built almost wholly on the principle of wave-particle duality with initially the Schrodinger equation and later the Schrodinger wave function dominating all aspects of electron-nucleus interactions. This includes the emission and absorption processes of the electron and also the actual propagation of light and electromagnetic radiation. But how sound are these theories and, in the final analysis, do they make sense? A fair idea of the soundness of the Standard Model theory of photon emission and absorption can be gained by posing a few questions. In the following passage questions are posed as numbered paragraphs, while the Standard Model answers are given in italics followed by citations from various authenticated quantum mechanics text books:

Q1. The present day ontology defining the electron is sadly out-dated, illogical and does not fit in with what is known. For instance, take the idea that electron emission and absorption is a one-off event. The electron cloud around the nucleus, absorbs an incoming photon and then at some undefined period of time, emits a photon. (How does one wave absorb another wave?).

Ans 1.The interaction between the electron wave-function and the electromagnetic wave (photon) is described by the interaction Hamiltonian. This term in the Hamiltonian represents the energy of interaction between the electron and the electromagnetic field.

References: Sakurai, J. J., & Napolitano, J. (2017). *Modern Quantum Mechanics* (2nd ed.). Cambridge University Press. :: Chapter 5: "Symmetry in Quantum Mechanics" and Chapter 7: "Time-Dependent Perturbation Theory."

Comment on the quantum mechanics viewpoint for Q 1.

While this mathematical explanation on how it is possible for one wave to absorb another wave might hold good and make sense to some, it doesn't explain the actual behaviour of photons as being emitted at the rate of hundreds of trillions per second. Believe it or not this is how fast things have been found to work in the present day. Nor does the explanation deal with the question of recoil or the conservation of energy and momentum in the classical sense.

Q2, Does the electron cloud comprise all the electrons in a multiple electron atom? What happens in multiple electron atoms how does this shift from one energy level to another work? Then after some indeterminate time, the electron cloud drops back down to its original level and the difference in energy is emitted as a photon. (How does an electron cloud emit an energetic photon without recoil, surely a breach of the conservation of energy laws?).

The quantum mechanics explanation is that:

Ans 2. The added energy from the absorbed photon results in the whole cloud moving to a higher energy state. In classical mechanics, the emission of a photon by an atom would cause the atom to recoil due to conservation of momentum. However, in quantum mechanics, the situation is different. The electron cloud's change in energy levels doesn't involve a point-like particle emitting a photon from a single location. Instead, it involves the entire probability distribution of the electron's position changing state.

References : Griffiths, D. J. (2018). Introduction to Quantum Mechanics (3rd ed.). Cambridge University Press. Chapter 9: "Identical Particles" and Chapter 10: "The Hydrogen Atom". Sakurai, J.J., and Jim Napolitano. Modern Quantum Mechanics. In Chapter 2, Section 2.4, "The Photon," the text explains how the concept of frequency is used in the quantization of the electromagnetic field to relate the energy of a photon to its wave properties.

Comments on the quantum mechanics viewpoint for Q 2.

Therefore the Standard model of quantum mechanics states that no recoil is present when one wave meets another wave and produces a photon, which in turn means that the laws of energy conservation turn a blind eye to what is a macro event.

The explanations put forward by quantum mechanics and buoyed up and supported by 'rigorous' mathematical models is supposedly an acceptable model of the atom and its interactions with light, but such explanations can and should be challenged.

Sooner or later the truth will have to be faced, namely that quantum mechanics is an out-dated theory striving desperately to defend indefensible theories. But do alternatives exist. Fortunately yes, new theories are coming to light that closely mirror classical physics while closely incorporating many quantum mechanics concepts and ideas, yet explain the same concepts at present dealt with by classical physics. In the next chapter some of these new ideas will be explored. This new theory of physics can be said to be a fusion of classical physics and quantum mechanics.

The aether

One of the reasons that the concept of an aether or medium through which light could travel was so popular with scientists of the late 19th and early 20th Centuries, was that it offered the perfect explanation for almost all of the properties of light. The aether explained how light travelled from point A to point B, it explained why the speed of light was constant regardless of the speed of the source or of the observer, it explained why light spread out directly as the square of the distance travelled and how its intensity varied according to the inverse square of the distance travelled. Yet, the findings of Maxwell, Einstein and Lorentz brought all discussion to a head. Maxwell introduced the concept of self-sustaining electric and magnetic fields, while Einstein, stated that the presence of electric and magnetic fields meant that the aether was redundant, finally Lorentz was able to give an explanation in mathematical form that explained why the speed of light was constant in the absence of an aether.

Aether

However, it was Max Planck who in 1900 made the path breaking discovery that light, which for centuries had been regarded as being smooth and continuous was in fact made up of tiny infinitesimal, discrete, indivisible packets of energy that he named quanta, who contributed the deciding factor. Once energy had been identified as being made up of discrete particles or quanta, the next logical step was to wonder if these little particles obeyed Galilean transformations or if they would behave as waves travelling through a medium. But this conjecture raised problems, if light were indeed made up of little discrete particles, how could it spread out according to the inverse square law. That being said, does any chance remain of reviving this ancient theory of an aether or medium that light travels through? Surprisingly, if one ignores the inherent, stubbornness and obtuseness of a large part of mankind, the odds for the revival of an aether like concept (not the luminiferous aether) are fairly good. After all, it was Max Planck who said:

"A new scientific truth does not triumph by convincing its opponents and making them see the light, but rather because its opponents eventually die, and a new generation grows up that is familiar with it."

— **Max Planck, Scientific Autobiography and Other Papers**

But, what would this new theory of the aether be? How could it be proved? Is it possible to detect this substance? More importantly what is the theory that it is trying to displace?

Special & General Relativity

The theory of the aether was replaced by Einstein's theory of Special Relativity, which is based on two postulates.

1) The first postulate of special relativity states that the speed of light remains constant in all frames of reference.

2) The second postulate of special relativity states that the laws of physics remain the same in all inertial frames of reference.

The Laws of physics remain the same in all inertial frames of reference:

Taking the second of these two postulates first, Einstein stipulated that the laws of physics remain constant in all frames of reference. In order to illustrate this concept imagine a card game with four players set up on a station platform and an identical card table set up on a train moving past that platform at a constant speed. It is found that the laws of physics are identical for both sets of card players. It is not possible to differentiate, without reference to outside references, which system is moving and which is still. Think of a boy bouncing a ball in the train and another boy bouncing a ball on the platform, in both cases the ball goes straight up and down. There is no difference between the two actions. The situation would be different if the train was accelerating or in a non-inertial frame of reference.

The speed of light remains constant in all frames of reference.

With regard to the first postulate that Einstein made namely that the speed of light was constant for all observers, he was never able to explain why this was so, he merely stated that this was the case. This is similar to the way in which Isaac Newton was able to describe gravity in detail but was never able to state the causative factor behind gravity. In trying to understand why this postulate involving the constancy of the speed of light proposed by Einstein is wrong, it must be understood that in nature, different substances travel in different ways. Solid objects obey the laws of Galilean transformations. For instance if there are two cars A and B, travelling away from each other; car A travelling at 60 kmh and car B travelling at 90 kmh we can say that they are moving away from each other with a relative speed of 90 kmh + 60 kmh or 150 kmh. If instead car A and car B are travelling in the same direction, we can say that car B is moving away from car A with a relative speed of 90 kmh − 60 kmh = 30 kmh. As can be seen Galilean transformations add and subtract, they are cumulative speeds.

How waves travel:

Waves travel in a manner that is different from the way in which solid objects travel. The speed of a wave depends solely upon the nature of the medium that the wave is travelling through. In the case of a wave, it does not matter if the source is moving or the destination is moving or even if any observer is moving towards or away from the wave. The speed of the wave will remain constant for all observers. Therefore, as long as the medium remains constant the speed of a wave travelling through that medium will also remain constant.

Einstein's take on the speed of light:

When he formulated his theory of special relativity Einstein was well aware of the two distinctions, namely the difference in between the way in which solids travel, obeying Galilean transformations and the way in which waves travel moving at a constant speed in a medium. He nevertheless treated light as a solid object because Max Planck had conclusively proven through experiments that light was made up of tiny, discrete packets of energy and was not continuous in the way that waves, moving in water may be thought of as being continuous. Einstein therefore treated light as a solid. There was a catch however. The problem that Einstein was faced with was this, although there was convincing proof that light was discrete in nature (i.e., particle like) it did not travel according to Galilean transformations but as a wave, in other words its speed remained constant. The manner in which Einstein explained this paradox is what special relativity is all about.

In order to make the speed of light constant to observers in all frames of reference, in spite of being particle like, Einstein stated that time must dilate or expand and that distances must contract. Since, distance depends on the observer's relative motion and is the product of time and speed, shorter time entails a shorter distance covered. The velocity of a particular object relative to an observer at rest is the proper length divided by the dilated time. By using this artifice (no other word comes to mind that is suitable) and using the Lorentz transformations formulated by the Dutch mathematician Henri Lorentz in order to explain why the aether was not detectable, Einstein was able to formulate a system whereby the speed of light was constant to all observers, regardless of whether they were moving or stationary.

Looked at in an objective manner this is an amazing assertion to make for it means that neither time nor space are fixed and immutable as they were hitherto thought to be. In hindsight it is amazing that special relativity came to be so widely accepted, although it should be pointed out that Einstein never received a Nobel prize for his formulation of special relativity because of opposition from the majority of eminent scientists of his time, including, Rutherford, the Curies, Ernest Gehrcke and Phillip Lenard. Therefore, it becomes apparent that although today both Einstein and special relativity are held sacrosanct by the scientific community, in reality, adequate grounds exist to question the validity of special relativity.

Returning to the concept of the aether, the very mention of an aether like substance has become anathema to most scientists, they will not even tolerate the mention of the existence of such a substance. Yet, it should be remembered that our very existence depends on the air in our atmosphere, the existence of which was not suspected, and was denied for hundreds if not thousands of years. To explain this new theory of the Aether, it is first necessary to examine the frequency of photons.

Frequency of photons:

One of the properties of the photon that has remained largely unexplained and vague in the Standard Model is the concept of the frequency of the photon. Frequency as applied to a single photon is very vaguely defined, with the general consensus being that only waves have frequency and that as applied to a single photon, the concept of frequency is a mathematical abstraction used solely to determine the photon's energy. However, modern technology has amply demonstrated that this ontology of photon frequency is dated. Take for instance the example of modern smart phones, they have the ability to process data at the rate of several Gigahertz per second. When this information is applied to the electron it is apparent that the electron, given its miniscule size and the infinitesimal sub-atomic distances involved in any oscillation; should be able to oscillate not at Gigahertz per second but on the order of hundreds of trillions of Hertz per second. In fact, for the electron not to be able to oscillate at such rates would be an anomaly. The invention and working of optical atomic clocks such as the Rubidium atomic clock, lends credence and absolute proof to the theory advanced in this paper that photon frequency is not a mathematical abstraction but that it has a physical existence. Therefore, when speaking of a photon frequency of 500 Terahertz, it means exactly that. Namely, that the electron is oscillating at the rate of five hundred trillion times per second and absorbing and emitting photons at that rate. Therefore, the matter to photon ratio is calculated at several hundreds of trillions of photons being formed per second for every particle of ordinary matter.

Photons at the time of the Big Bang:

If now, one thinks back to the time of the Big Bang one would have to take into consideration that during the process when primordial matter was being formed, it would have been accompanied by the creation of photons (light) in the ratio of hundreds of trillions of photons per second for every particle of matter that was formed. While it is known that the Big Bang was not an explosion in the conventional sense but more a rapid expansion, there is almost a universal consensus that the Big Bang would have been accompanied by light, although not in the form we know of today. This process of the creation of matter in the early Universe would have gone on for thousands or even millions of years, during the creation of the early Universe. The point to consider is what happened to all these early photons that were being created in unimaginable numbers? They could not by definition have crossed over the edges of the Universe, since by definition nothing was there. So, they began to accumulate within the early Universe, filling every part of it. The dipole structure of these photons (See Figures 3, 4 & 5) meant that they were able to connect together to form linked networks. As the Universe continued to expand the linked virtual photon network expanded with it. Eventually the whole of the universe was completely permeated by this network of linked photons that formed a continuous background to the Universe, permeating every part of the Universe including in all matter. Eventually light began to propagate in the manner that is familiar to us today.

Figure 3.

Figure 4 Photons linked in series

Photons connected in parallel

+

Figure 5. Photons linked in parallel

The continued expansion of the Universe, resulted in the virtual photon network losing energy, until eventually each individual virtual photon had an energy of about 10^{-51} J. This low energy meant that no atom, which are very specific about which energies they can interact with, could possibly interact with photons of such low energy as 10^{-51} J. The virtual photons of the virtual photon field, could go straight through the atom without experiencing the slightest interaction and vice-versa. Therefore, the network of virtual photons that permeated the whole of the Universe and formed the background fabric of the Universe was absolutely permeable to matter. It is estimated that virtual photons as described have such low interaction with matter that they could pass through a block of lead a light year thick without experiencing any interaction.

Therefore this interpretation of the Aether, is far removed from the concept of a luminiferous aether prevailing at the end of the nineteenth Century. The luminiferous aether possessed such contrasting properties; having to be simultaneously permeable to matter and possessing a rigidity many times that of steel, that it was an insupportable theory. By contrast the aether that is described in this Chapter, more closely resembles the classical definition of a field, in physics. In classical physics a field can be described as a region in which each point has a physical quantity associated with it. The quantity could be a number, as in the case of a scalar field such as the Higgs field, or it could be a vector, as in the case of fields such as the gravitational field, which are associated with a force. The question that is uppermost at this moment is how did the existence of the aether come about?

Photon Frequency:

To gain some idea of how virtual photons came to permeate the entire Universe, it is first necessary to examine in detail another of the out dated concepts of quantum mechanics. This is the concept of the frequency of a photon. According to quantum mechanics:

"Photons don't have a frequency. Frequency is a wave property and is a property that applies only when you look at light as a wave. A photon has an energy, which is related to, but different from, frequency. So we sometimes use the term 'the frequency of a photon' but really as a

short hand for "the frequency of the wave that will manifest later as a photon".

Dr. Phillip Freeman on Quora: https://www.quora.com/How-are-photons-frequencies-and-energy-related.

Another way of stating this is by $f = \frac{e}{h}$ this means that if you know the energy of the photon it is possible to determine its frequency by dividing the photon energy by h planck's constant. Whatever the answer might be it is fairly obvious that the quantum mechanics concept of what the frequency of a single photon might be is shrouded in obscurity; due to the wave-particle duality there can be no definite answer to the question.

This presents a problem when looked at from a purely mechanical point of view. The problem arises as follows, if one looks at a modern day smart phone one is aware that it is ***processing*** data at the rate of several gigabits per second. Consider what the term ***processing*** denotes, it means taking input data, evaluating it and storing or outputting the result. The electron is miniscule in size and the distances over which it has to oscillate are even smaller. It is then only natural that the electron should oscillate at frequencies of several hundreds of terahertz and emit photons at that rate. In fact not to do so would be odd. Gestalt Aether Theory, therefore, takes the concept of photon frequency away from the fuzzy abstract notions of quantum mechanics where it is an abstract mathematical

112

property of the photon and puts it on a sound practical basis. Therefore when we talk of an electron emitting photons with a frequency of 600 THz it means exactly that. The electron is emitting photons at the prodigious rate of 600,000,000,000,000 photons per second. In what direction are these photons emitted? These photons are emitted in a single direction as a line of photons of the same frequency, wavelength and energy. Why? If one looks at the physics behind the emission of a photon, it is apparent that the electron absorbs energy and mediates its energy by emitting that energy in the form of a photon, the process involves the force of recoil. To cope with these forces of recoil the electron travels toward the massive nucleus, recoils off the nucleus and again emits this energy at the exact same position as before but in the opposite direction, hence photons are emitted as a line of photons in a specific direction. Proof that this is indeed the case can be seen in the working of any atomic clock. Frequency is directly related to the rate of emission of photons by the electron or to put it another way by the electron's rate of oscillation. There exists ample evidence that this theory of how photon frequency is linked to rate of emission or oscillation of the electron is correct in the working of the new rubidium optical atomic clocks.

We are now in a position to understand the concept of a Universe completely submerged or steeped in a sea of virtual photons. Having established that photons are emitted and absorbed at phenomenal rates, it is time to return to the first moments of the Big Bang that is thought to have occurred out of a singularity more than 13.7 billion years ago! Although it is generally accepted that the Big Bang was not a bang in the accepted sense of an explosion but that it involved a rapid inflation or expansion, opinions is almost unanimous that light must have been present also. It is almost a corollary that if matter was present light (or rather photons) must have been present also, although light might not at this early stage of the Universe have been able to propagate. The question is what happened to all that light, look at the ratio; for every atom of matter, photons were being emitted at the rate of hundreds of Terahertz per second (10^{14}) per second. What happened to those photons? Emitted not over seconds but hundreds of thousands, even millions of years? They could not have escaped over the borders of the Universe because by definition nothing exists outside the Universe, so they must have been contained in some way within the Universe. Don't confuse this with CMBR (Cosmic microwave background radiation) for one thing Gestalt Aether Theory does not hold that optical photons can be red shifted, no matter what the circumstances, to the extent that they transform into microwaves.

In support of this statement we are now in a position to look at light from a little more than 13 billion years ago (i.e., a source that is 13 billion light years distant). If this is indeed the case what happened to the light that was created at that time along with the heat? It is presumed that if light could not escape over the borders of the Universe the scenario of a linked net- work of photons that permeate the whole of the Universe is a very plausible theory. If one examines the model for the structure of the photon proposed earlier, it is possible to see that these photons can link together not only vertically end to end but also laterally. When photons are linked in this manner their energy is shared. The theory is that as the Universe continued to expand the vast numbers of photons present formed a linked network. As time passed and the Universe continued to expand the network of linked photons expanded with the Universe filling every part of it. As the expansion of the Universe proceeded, the individual energy of the photons was shared out among the network of linked photons, till eventually they reached such a low state of energy (about 10^{-51}J) that they for all practical purposes ceased to exist! Because of this extremely low energy it was possible for this linked network of photons that permeated the entire Universe, to exist for practically ever, with life-times similar to that of the proton or the electron.

The Heisenberg Uncertainty Principle as it applies to time and energy:

This process can be attributed to the Heisenberg Uncertainty Principle as it relates to time and energy:

115

$\Delta e \ \Delta t > h$ Given that energy $= 10^{-51} J$ and $h = 6.6$ x 10^{-34}

$(6.62 \ x \ 10^{-34}) / (\ 1.6 \ x \ 10^{-19}) = 4.13 \ x \ 10^{-15}$ s

The result of this calculation show that 'virtual photons with an energy of 10^{-51} J could exist for twenty billion years. The HUP states that if an event happens in an extremely short period of time $< 10^{-15}$ s or possesses very small energies, it can be ignored by the laws of energy conservation which rule all processes at the macro level.

These 'virtual photons' of the virtual photon aether were almost stationary dipole particles with which, because of their extremely low energy, no matter would or could interact, thus they could travel through matter as if it did not exist and vice versa. Eventually when all of the Universe was full of this linked network of photons, light began to propagate in the way that is familiar to us today. When an electron within an atom emits a real photon the photons of the virtual photon aether, line up forming a line whose ends rest on the shoulders of infinity and the energy of the real photon travels along this line of linked photons.

This explanation of how an aether like medium formed, consisting of extremely low energy $(10^{-51}$ J) photons that permeated every part of the Universe and yet was absolutely permeable to all forms of matter and would allow massive objects such as the planets the sun, the moon and the stars to pass through it without the slightest opposition, explains everything that was ever known about light. The aether does not interact with matter because no atom could possibly need or utilise such low energies, thus virtual photons of the aether pass through all atoms as if they did not exist. The existence of such an aether also explains in easily understandable terms, how light can propagate according to the inverse square law and still enable individual photons to maintain the same energies that they possessed at the time they were created. This Aether model explains how and why the speed of light is constant and invariant and unaffected by the motion of either the source or the observer or of both.   This is a perfectly modelled system for light and all of its properties; it makes sense as the quantum mechanics model for light and electromagnetic radiation fails to do. The quantum mechanics view of light is that it travels from point A to Point B as an abstract wave-function based on the Schrodinger wave equation, this in turn means (or should mean because of the multiple particle nature of light) that the wave function is travelling through multiple dimensions, it also does not exist as something real as Richard Feynman had stated. The light only becomes real when it is detected at point B, when the light is detected it becomes real once more and the wave function collapses. By contrast the Gestalt Aether Theory on the propagation of light is lucid and its reasoning is easy to follow.

The aether described herein is made up of infinitesimal discrete electric dipole points that can easily orientate themselves in the direction of an electric field. Therefore when a real photon is emitted by an electron, the virtual photons of the 'virtual photon field' that permeates every part of the Universe, and are normally oriented at random, align themselves in the direction of propagation of the emitted photon forming into a line whose ends rest on the shoulders of infinity, and the energy of the real photon travels along this line of aligned virtual photons. It should be pointed out here that as 'real' photons are emitted in the same direction in a line of identical discrete units of energy, frequency and wavelength, that the line of aligned virtual photons forms a ray of light that is travelling rectilinearly, The emitted 'real' photon and the 'virtual' photon of the virtual photon field are identical in physical size and structure with the exception of the energy they carry, As the energy of the real photon that is emitted travels along the line of linked virtual photons, it experiences no net loss in energy since the virtual photons have their own base energy of about 10^{-51} J , that they have to maintain.

Heisenberg's Uncertainty Principle and the Laws of Thermodynamics:

Another explanation for the longevity of virtual photons of the virtual photon field can be found in Heisenberg's uncertainty principle as it has to do with energy and time $\Delta E \, \Delta T \geq h$. This equation states that if an interaction takes place over a very short period of time, then the energy that it possesses becomes arbitrary, similarly if an interaction has extremely low energy the time over which it can exist becomes arbitrary. As for instance:

Here it is interesting to hypothesize on how the virtual photons formed at the time of the Big Bang, were able to survive practically untouched for billions of years. Of course other particles such as electrons and quarks have also survived over the same time scales. As far as virtual photons goes, it is supposed that their existence dates from the epoch of the Big Bang and therefore is closely linked with the three laws of Thermodynamics. For instance according to the first law of thermodynamics, matter can neither be created no destroyed, therefore these early photons represent part of the initial energy that went into forming of the Universe and hence are an integral part of the Universe forming the background fabric of the Universe. According to the second law of thermodynamics, energy or heat, moves in a one way direction. Heat always moves from a hot region to a cold region and never the other way around. Lastly the third law of thermodynamics states that the entropy of a closed system at thermodynamic equilibrium approaches a constant value when its temperature is zero.

$\Delta E \, \Delta T \geq h$

Since the value of E is known to be 10^{-51} J and Planck's constant is

119

$$6.6 \times 10^{-34}$$

an approximate idea can be had of the times involved: $(6.6 \times 10^{-34})/(10^{-51}) = 6.6 \times 10^{17}$ seconds, if we convert seconds to years = $(6.62 \times 10^{17})/(31557600) = 2.2 \times 10^{10}$ years. In other words photons with such low energies (10^{-51} J) can have an age larger than that of the Universe.

The advantages that such an aether possesses in explaining all the properties of light and electromagnetic radiation is exceptional, there is no aspect of light that cannot be explained using the aether theory including the property of reflection, propagation of light according to the inverse square law, preserving of energy intact from time of emission to time of absorption, rectilinear propagation of light etc., All of these properties of light require elaborate, very complicated, mathematically abstract explanations in quantum mechanics. These explanations for the nature of light are accompanied by an oppressive dependence on esoteric phenomena.

Where is the aether, why can't we see or detect it?

Max Born

The answer to this question is that atoms will only interact with photons of certain energies, if these energies are not available, the photon will pass straight through the atom without experiencing any interaction. This is why objects possess different colours, they only interact with photons of specific energy, hence the presence of objects with different colours. In the case of virtual photons with their extremely low energy of 10^{-51} J it is inconceivable that any atom could or would interact with photons of such low energy. This is why the aether is undetectable, it has extremely low to zero interaction with matter. When considering this non-existent interactivity of the aether with matter, it is interesting to remember Max Born's, often regarded as the father of quantum mechanics, tongue in cheek comment about the aether:

"One obvious objection to the hypothesis of an elastic Aether (Space) arises from the necessity of ascribing to it the great rigidity it must have to account for the high velocity of Waves. Such a substance would necessarily offer resistance to the motion of heavenly

bodies, particularly to that of planets. Astronomy has never detected departures from Newton's Laws of Motion that would point to such a resistance." (Max Born, on Quantum Theory, 1924)

The aether herein described, overcomes all of Max Born's objections to the existence of an aether, for one thing the great rigidity that he speaks of is not required. The question still remains as to where this elusive aether is. The answer is staggering in its implications, the virtual aether field that we refer to is nothing less than Dark Matter. Yes, what we formerly took to be aether is in fact Dark Matter. Consider the evidence, Dark Matter accounts for 85% to 95% of all matter in the Universe, since this is the case, the odds are heavily weighted that our own solar system falls within that 85% - 95% of space occupied by Dark Matter. This means that what we took to be the aether, was in fact Dark Matter, both of these substances possess identical properties. Next we will consider a very important aspect on the propagation of light.

The Gestalt Aether Theory On The Propagation of Light:

The term 'light' is deliberately used in the sub-title instead of 'electromagnetic radiation' since according to Gestalt Aether Theory there is a small but significant difference in the manner of production and propagation of high energy, high frequency photons like visible light and x-rays and gamma rays and relatively low frequency, long wavelength, low energy radio-waves.

In order to introduce a new theory or concept it is first necessary to review how the old system worked. Feynman's is generally considered to be the ultimate expert on Quantum Electrodynamics or QED theory so here is his theory of light propagation:

"Feynman explains that in QED light sources produce not physical particles or waves, but wavelike "probability amplitudes" that propagate at c in space (not superluminally). The amplitudes spread in all directions and superpose (interfere) just as real light waves do according to the Huygens-Fresnel principle: by spherical wavelets from every portion of the wave front. Feynman restates this principle as light "has a nearly equal chance of going on any path". As they propagate in space, the probability amplitudes shrink according to the inverse square law and rotate in space according to their frequency ("shrinks and turns"). Adding up all the resultant arrows for all the possible paths light may travel to the receiver renders a final amplitude arrow. Squaring this arrow yields the probability that a detectable light-matter interaction will be observed. Where the probability amplitudes superpose constructively is where events (e.g. photomultiplier counts) are more likely to occur; where they superpose destructively is where events are less likely to occur.

Feynman admits that the wave theory of light can account for all the phenomena modelled by QED when the light is intense; but insists that "wave theory cannot explain how the (photomultiplier) detector makes equally loud clicks as the light gets dimmer."

http://henrylindner.net/Writings/PhysEssSpacePhysics2.pdf

Reading the above it is immediately possible to realise that quantum mechanics (at least as it has to do with wave-particle duality) has produced a clouded and confused narrative about light, which apart from complicated maths based almost entirely on imaginary numbers (multiple dimensions) does not offer much of an explanation for anything. Quantum theory has never been adequately able to explain the propagation of light according to the inverse square law. This is a huge lacuna considering that 400 years have passed since Newton and Huygens and 150 years since Maxwell. We should, by this time, have had a beautiful theory on how light propagates, we don't. Gestalt Aether Theory offers one such beautiful and unified theory.

But what is the actual mechanism by which light propagates in Gestalt Aether Theory? Light propagates as follows. The 'virtual photons' of the aether which incidentally resemble a field of to all purposes stationary electric di-poles, are oriented at random until a real photon is emitted by an electron, when this happens the 'virtual photons' of the aether, in the line of propagation of the emitted real photon align in the direction of propagation of the real photon forming themselves into a line whose ends rest on the shoulders of infinity. The energy of the real photon then travels along this line of virtual photons. However, and this is of significance, as the energy of the line of real photons travels along the line of aligned 'virtual photons' it is dispersed not only forward but also laterally so that the energy from the line of real photons spreads out in a cone shape from its point of origin. All of the area covered by this shape is filled with the energy of the original line of photons being emitted at the source. In order to understand how this dispersion takes place it is necessary to understand the concept of ***promotion*** wherein 'virtual photons' adjacent to and in contact with the energy of the line of real photons are promoted to the status of real photons through a transfer of energy. In order to understand how this happens, look at the lead photon, it has been emitted at the head of a line of connected photons and it is moving away (its energy) is moving away at the speed of light. Each photon in this line of connected photons possesses identical energies. When a photon in this line of connected photons comes into contact with virtual photons, the virtual photons will be aligned in the direction of propagation of the real photon, this contact results in an energy transfer not only forward but laterally also. This means that a virtual photon moving beside the real photon and in contact with it will acquire all of the energy from the real photon and the energy of the real photon that had transferred its energy will be immediately

125

replenished from the line of photons behind it. This is why light, as it propagates, follows the inverse square law of dispersion. In incoherent light every receptive electron in the receptive substance will be emitting light at the rate of whatever frequency light is exciting it (i.e. the electron) this results in lines of photons being emitted in different directions all travelling at the speed of light and all spreading out in keeping with the inverse square law. The direction in which these lines of photons are emitted depends on the laws of reflection as elucidated in classical physics. However, since on average a cubic centimetre of solid material contains about 10^{22} atoms the lines of photons emanating from the excited material are virtually innumerable. Thus as each line of photons is emitted by an electron it comes in touch with neighbouring 'virtual photons' surrounding it and it passes its energy to these adjacent photons, so each real photon passes all of its energy to its neighbouring 'virtual photon' which is immediately *promoted* to a real photon which in turn can pass its energy along to its adjacent 'virtual photons ' and so on. The energy of the real photon which has passed on its energy is almost instantly topped up or replenished from the line of real photons behind it. It is important to understand how the transfer of energy takes place during the course of promotion of the virtual photon to a real photon. The real photons energy is fixed by its configuration, its configuration at the time it was emitted determines its energy, as a result it can only pass on or

126

receive whole amounts of this energy, nothing less and nothing more. A virtual photon that has been promoted to the status of real photon, maintains it original energy level of 10^{-51} and the energy of the real photon is superposed on this base energy and hence is always passed on in whole units. This is the manner in which light disperses according to the inverse square law. It can now be seen that the inverse square law is a function of the ratio of how many virtual photons come into contact with a real photon. If the real photon is exposed on all sides then it is simultaneously accessible to six virtual photons that will each in turn acquire energy and be promoted to a real photon. However because of the huge number of electrons that are simultaneously emitting photons, the number of virtual photons that can interact with a real photon is limited. It can be seen that the inverse square law is an approximation of this process of promotion of virtual photons. When the source electron stops emitting and there is no longer any energy, all of the promoted photons lose their energy and turn into 'virtual photons', fading back into the virtual photon aether.

Note how natural this version of the propagation and dispersion of light is, it is not the physical real photon that travels but only its energy. In this way the propagation of light is brought into line with the propagation of *all* other manner of waves such as water waves sound waves etc., with one vital distinction. This distinction is that for as long as light is being emitted, (i.e., for as long as the photon is being excited and emitting photons) every photon in the propagating *'wave* front' *retains* the same *energy* as the *originally* emitted photon.

As the light moves further from its origin the intensity of the light varies as the inverse of the square of the distance from the source. However, the energy of each individual photon remains intact, thus if it is monochromatic blue-green light with a frequency of 600 THz that is being emitted then the energy of each individual photon would be equal to

$(3\times10^8)\times (6.62607004\times10^{-34})/(5\times10^{-7} = 2.483$ eV.

Thus at every point, considering that the monochromatic light has been travelling 100 Km, on a 10 billion square metre front, the energy would be 2.48 eV. Energy from the original line of photons being pumped out by an electron is being dispersed, even as it moves forward at the speed of light, through all the virtual photons immediately surrounding it laterally, these in turn are promoted to real photons, possessing an energy identical to that of the originally emitted photon and passing on that energy to neighbouring virtual photons, *promoting* them to real photons with identical energies to the original emitted photon, while also replenishing its own energy, from the original line of real photons.

Thus at 2 metres distance from the origin, the original line of photons, taking as an example the same 500 nm wavelength, that is being pumped out at the rate of 6×10^{14} per second by an individual oscillating electron, has shared its energy out between three virtual photons, promoting them to real photons and making a total of four real photons, at 3 metres distance from the origin the 4 real photons would have become 9 real photons, the energy in the line behind the original photon consisting of 6×10^{14} photons per second being passed onto the adjoining virtual photons and 'promoting' them to the level of real photons, these photons in turn are passing on their energy and instantly replenishing it so that wave front consists of photons of a uniform energy.

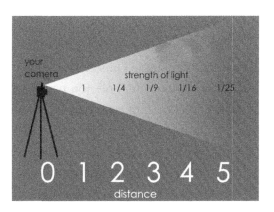

Inverse square law

Thus at 2 metres distance from the origin, the original line of photons, taking as an example the same

500 nm wavelength, that is being pumped out at the rate of 6×10^{14} per second by an individual oscillating electron, has shared its energy out between three virtual photons, promoting them to real photons and making a total of four real photons, at 3 metres distance from the origin the 4 real photons would have become 9 real photons, the energy in the line behind the original photon consisting of 6×10^{14} photons per second being passed onto the adjoining virtual photons and 'promoting' them to the level of real photons. At 6 metres from the source there would be 36 'real' photons and so on and at 9 m there would be 81 real photons and at 100 m there would be 10,000 real photons and so on. It should be noted that this is just a simulation using a single line of photons. In reality at a distance of a 100 metres from the source there would a huge number of linked real photons, this is a necessary corollary of simplifying the depiction. Thus as the distance from the origin increases, the original photon promotes virtual photons to real photons, in a series that increases in number exponentially; as in the series, 2, 4, 9, 16, 25, 36, 49 etc., Similarly, as the distance increases, the energy in the line of photons behind the real photon eventually reaches its limit (i.e., the electron stops emitting) this means that the core energy of the photon is now being used, very soon, the light fades and all of the real photons that have been created in this process, turn back into virtual photons. This process also explains how aside from spreading out the intensity of the photons is reduced

inversely in proportion to the square of the distance travelled because there are less photons present behind the photons forming the wave front resulting in a reduction in intensity.

Description of photon propagation through Universal photon aether:

Interaction of Real Photons with a Sea of Virtual Photons

Both real photons and virtual photons are electric dipoles. Real photons are emitted as a line of electric dipoles aligned end to end. When a real photon or a line of real photons is emitted, the extremely low energy (10^{-51} J) electric dipoles in the sea of virtual photons that permeate the whole of the Universe and form the background fabric of the Universe, align themselves end to end along the path of propagation. This alignment facilitates the transfer of energy through the medium of virtual photons.

As the line of real photons travels in a straight line at the speed of light through the line of aligned virtual photons, they encounter adjacent virtual photons. The energy of the real photon is shared in two primary ways: The energy is passed along the line of aligned virtual photons. This energy transfer continues along the path of propagation, maintaining the alignment and energy flow. In addition to the vertical transfer, real photons also share energy laterally with adjacent virtual photons. This lateral interaction ensures that the energy of the real photons spreads into the surrounding virtual photon sea, promoting more virtual photons to real photon status and expanding the wave front.

The alignment and interaction of virtual photons with the real photon line lead to the formation of a wave front. The wave front expands as energy is shared both along the line of real photons and across the virtual photon sea, reflecting the spreading nature of the propagated energy. The energy distribution follows the inverse square law, where the area of the wave front increases proportionally to the square of the distance travelled. This ensures that the energy from the line of real photons continues to propagate and spread effectively.

In summary, when a line of real photons is emitted, it aligns the virtual photons end to end in its path. As the real photons travel, they share their energy vertically along this alignment and laterally with adjacent virtual photons, contributing to the expansion of the wave front and the continuous propagation of energy.

To describe the process mathematically, let's consider the following aspects:

A line of real photons, represented by N photons in a line, where N is very large (hundreds of trillions per second). Each real photon has energy E. The energy of the line of real photons propagates through a medium of virtual photons. Virtual photons along the direction of propagation of the line of aligned photons, align end to end along the line of propagation of the real photons, and the energy of the real photons travels along this line of aligned virtual photons at the speed of light promoting them to the status of real photons. The virtual photons in the virtual photon sea that come into contact with the line of real aligned photons are also promoted to the status of real photons, possessing the same energy and characteristics, and also align themselves laterally and vertically in the direction of propagation of the line of real photons.

The transfer of energy takes place in two orientations, vertically (i.e., in a straight line) and laterally (i.e., sideways.). The area of the resulting wave front A is proportional to the square of the distance r travelled, in accordance with the inverse square law.

Energy Transfer Along the Line of connected photons :

Let E_i be the energy of each real photon in the line. The total energy E_{total} of the line of N photons is:

$E_{total} = N . E$

As each virtual photon is promoted to a real photon it aligns with the line of real photons, the energy transfer per virtual photon E_v is:

$E_v = E/N$

As real photons move, they continuously promote adjacent virtual photons, maintaining the alignment. The energy distribution follows the inverse square law. The area A of the wave front at a distance r from the source is:

$$A = 4\,\pi r^2$$

The energy density ρ at distance r is inversely proportional to r^2 since there is no longer a huge number of real photons behind each real photon in the wave front.

$$\rho(r) = E_{total} / A = N\,.\,E/\,4\pi\,r^2$$

The energy of each real photon is continuously replenished by hundreds of trillions of photons behind it. The replenishment rate R can be represented as:

R = constant, 10^{14} photons per second

The energy E_v of each promoted virtual photon continues to propagate along the line.

While the lateral energy transfer to adjacent virtual photons ensures that energy is distributed across adjacent virtual photons, expanding the wave front.

Analogy with Water Molecules:

Each water molecule oscillates up and down, transferring energy to adjacent molecules. The overall wave propagates through the medium of the water consisting of water molecules. Similarly, each real photon transfers its energy to virtual photons in its path, promoting them to real photons. Promoted virtual photons further transfer energy to adjacent virtual photons. The area of the wave front increases proportionally to the square of the distance travelled, in keeping with the inverse square law. The energy is constantly replenished by the line of real photons in the line of connected real photons that are being emitted at the rate of 10^{14} per second thereby maintaining a continuous propagation front.

Interpretation:

In this model of the propagation of light, a real photon propagates through a sea of virtual photons by promoting virtual photons to real photons and transferring energy locally. Promoted virtual photons further pass on their energy to adjacent virtual photons, spreading the wave front. The area of the wave front increases according to the square of the distance travelled, in line with the inverse square law. The continuous production of photons ensures that energy is constantly replenished, allowing for persistent propagation over time.

Thus light is created it travels through light, in turn creating more light, and in the end returns back into light. This is exactly similar to the manner in which a wave disperses its energy. It is possible to see how simply and naturally dispersion is explained using the aether model. Light waves spread out naturally and in the same manner as when a stone is dropped into a still pool of water. To those who have read this theory carefully it will be apparent how effortlessly and exactly the manner of propagation of light has been explained. It has been explained in plain language how light is able to travel as a wave, obey the inverse square law of dispersion and at the same time retain its individual identity in the form of its original energy. No mathematical abstract wave function, no disembodiment and travelling through unworldly dimensions, no being in two places at once. Here is an extremely realistic view of the propagation of light that continues the tradition of Newton, Descartes, Lorentz, Maxwell, Poincare and others who passionately pursued the existence of an aether simply because it answered all possible questions about light and its propagation as both quantum mechanics and relativity could not.

Definitive proof of the existence of an aether

The proof that the aether exists and that it permeates the whole of the Universe, can be had by asking a very simple question. Why, does light as it travels from point A to point B follow the inverse square law, namely spreading out over an area equal to the square of the distance travelled and losing intensity according to the inverse of the square of the distance travelled? If electromagnetic radiation simply travelled in a straight line from point A to point b with nothing to impede its progress, there would be no overall reduction in intensity. Why, then with nothing to obstruct it, does electromagnetic radiation follow the inverse square law. Taking the case of sound waves travelling in air, it is found that sound, because it is travelling through a medium spreads out according to the inverse square law simply due to its interaction with the medium it is travelling through, namely air. Waves in water also spread out in a similar fashion in keeping with the inverse square law. But light we are told is not travelling through a medium, light does not need a medium to propagate, it should therefore travel in a straight line, similar to the way in which a bullet travels. Since, this is the case and given the circumstance that there is nothing in its way to impede its progress, why does light spread out following the inverse square law? Taking into account wave-particle duality, if light spreads out because it is travelling as a wave, what is the nature of the medium it is travelling through? Do electric and magnetic fields result in this kind of spreading out? Take as an example the Voyager 1 space craft which at present is 24 billion kilometres away, in deep space. The density of matter at this distance is on the order of one atom per cubic metre of space.

To understand what this means, imagine that the size of the atom were increased to the size of a 1 centimetre steel ball bearing. This would mean that the size of the square that it occupies is $(10^{-2})/(10^{-10}) = 10^8$ m or a square with sides of 100,000 km!

This means that there is literally nothing to disrupt the signal and the signal should be able to travel from its present location 24 billion kilometres (24,000,000,000 kilometers) distant without significant loss in intensity, the fact that its signal strength is reduced from an initial 23 W (add antenna gain of 30 DBi meaning a signal strength comparable to 23,000 W) to a mere 2.4 x 10^{18} W at the receiving antenna, shows that the transmission is following the inverse square law.   The question is why? If there is nothing with which to interact with, the signal should not spread out, this is just common sense. Thus the very fact that electromagnetic radiation follows the inverse square law, is definitive proof of the existence of an aether like medium through which light and electromagnetic radiation propagate.

Chapter 4: Electricity and Radio Waves

It is a well-known paradox in physics that a theory need not be perfect in order to work. An approximation will work as well as a more accurate model. This is especially true of physics in the modern day. Numerous theoretical models exist in explanation of various physical phenomena that are almost never used to calculate practical problems. For instance one never sees radio engineers trying to solve problems to do with radio transmission using quantum electrodynamics or quantum field theory, they inevitably use the simpler model developed by James Clerk Maxwell, even though this model might not represent the problem as accurately as the QED model. Similarly it is extremely rare to see the NASA engineer who uses Einstein's General Relativity to plot the course of a satellite around the solar system, instead the theory of gravity developed by Sir Isaac

Newton is almost universally used. The theory of electricity also faces similar scenarios, in fact during the past one hundred years or so since some understanding of the process has been attained there have been more than four different theories, each of which worked perfectly until its faults became known. Similarly the present Bloch wave-function model of electricity also works but which nevertheless, is wrong.

Even in the short time of about one and a quarter centuries, during which man has become proficient in the use of electricity, several theories have been formed on how electricity behaves, each new theory displacing a former theory. The very first theory of electricity could be said to have been formulated by Henri Lorentz and Paul Drude. Their theory was based on classical physics. The Lorentz- Drude model of electricity suggested that conductors such as metals contain a great number of free electrons. The positive ions are fixed at their sites. (This Positive Ion Arrangement is called lattice). In an open circuit, the negative ions (electrons) move at random in the lattice. Henri Lorentz had some interesting ideas about the electron. The Lorentz oscillator model, also known as the Drude-Lorentz oscillator model, involves modeling an electron as a driven damped harmonic oscillator. In this model the electron is connected to the nucleus via a hypothetical spring with a spring constant. The driving force is the oscillating electric field. However, the Drude-Lorentz model of electricity although it worked and is still used as a rough approximation today, failed in other ways. 1) It failed to explain the electric specific heat and the specific heat capacity of metals. 2) It failed to explain the superconducting properties of metals. 3) It failed to explain new phenomena like the photo-electric effect, the Compton effect and Black body radiation, etc.

The Drude Lorentz model of electricity was replaced by the Arnold Sommerfeld model of electricity. The quantum free electron theory is a model which is much used in solid state physics, to describe the behavior of electrons in a metal. It assumes that the electrons in a metal are free to move throughout the material, not bound to individual atoms. The key principles of the quantum free electron theory are the following points:

1) Electrons are treated as non-interacting, free particles that can move throughout the metal.

2) Electrons occupy discrete energy levels allowed by the infinite potential well model, with the energy levels given by the formula $E = (h \times n^2) / (8m \times L^2)$, where h is Planck's constant, n is the energy level, m is the electron mass, and L is the dimension of the metal.

At low temperatures, the electrons fill up the available energy levels according to the Pauli exclusion principle, with no more than two electrons (with opposite spins) occupying each level. The electrical and thermal properties of metals, such as electrical conductivity and heat capacity, can be explained by applying the principles of quantum mechanics to the free electrons in the metal.

The quantum free electron theory provides a good basic model for an understanding of many of the fundamental properties of metals, though it does have some limitations and neglects factors like electron-electron interactions. It laid the groundwork for more advanced theories of electrons in solids.

Finally the theory of electricity in use today is called the Zone theory (or Band Theory) and was invented by Felix Bloch. Bloch's Theorem is a foundational principle that explains the wave-like properties of electrons in a clear solid. This theorem, integral to quantum mechanics and the study of solid state materials, states that electrons in a periodic lattice behave as waves described by wave-functions known as Bloch functions. Bloch's Theorem is a foundational principle that explains the wave-like properties of electrons in a clear solid. This theorem, integral to quantum mechanics and the study of solid state materials, states that electrons in a periodic lattice behave as waves described by wave-functions known as Bloch functions. What is the Bloch state wave-function?

The Bloch state wave-function describes electrons in periodic potentials. It's periodic and represented by Bloch waves within the Brillouin zone, crucial for understanding solid-state physics.

143

The present theory of electricity based on the Gestalt Aether Theory, states that all of these models for the conduction of electricity are wrong for a very fundamental reason. These theories work on the assumption that it is the electron that is the fundamental charge carrier, although Bloch's band theory does suggest that electrical energy is carried along from particle to particle by a wave function. This theory although superficially explaining the phenomenon of electricity falls short in explaining associated phenomena such as the lines of force around an electrical conductor although a trivial explanation is put forward, and for an accurate account as to how exact amounts of electricity are transported and delivered.

The Gestalt Aether Theory of Electricity:

According to Gestalt Aether Theory, it is not electrons that are the fundamental charge carriers but photons. This is a fairly logical explanation since in every other known instance of electron energy exchange the exchange of energy is always mediated by photons. Why then did quantum mechanics reject this solution and opt for the more complicated wave-function theory proposed by Bloch? The answer ironically, is that the theory that electrical energy was carried by photons was rejected is that in order for a free electron to emit a photon, there has to be some mechanism present to absorb the resultant force of recoil. In the case of a bound electron within an atom,

144

the recoil force is absorbed by the massive nucleus in a free electron, this mechanism is not present. But why was the quantum mechanics explanation for the rejection of photons as charge carriers accepted? Apart from the fact that the photon is a neutral particle and was therefore thought to be incapable of conveying electrical energy, it was not possible to adopt photons as fundamental charge carriers due to the fact that for a free electron to emit a photon would violate the laws of conservation and momentum of classical physics. This is ironical because in the present theory of quantum mechanics; electrons within the atom do precisely that: emit photons without recoil taking place. Quantum mechanics manages to achieve a no recoil situation by treating both the electron and the photon as waves. The electron is represented by a wave-function and the photon is represented as an electromagnetic wave. Hence no recoil takes place. Although, the question of how one wave, the electron wave-function, can absorb another wave the electromagnetic wave representing the photon and then somehow contort itself to emit a photon with a precise energy value frequency etc., requires not only some very dexterous mathematics involving Hamiltonian exchange operators but also some extraordinary use of the imagination.

The conduction photon

Gestalt Aether Theory involves the introduction of the concept of limiting the size of the largest size photon that an electron can emit. This makes sense, considering that the classical radius of the electron is 2.82 x 10^{-15} m. Also when one considers radio-waves which possess properties identical to the photon and can reach wave-lengths in excess of 5 x 10^6 m. it becomes obvious that a tiny electron could not by any stretch of the imagination produce such massive photons.

A key concept in the Gestalt Aether theory of light is that of limiting the size of the longest wave-length photon that an electron can emit. Present theories hold that low energy electromagnetic radiation like radio waves are emitted by reason of the jiggling or oscillation of atoms and ions within an electrical conductor (lattice vibrations), while high energy electromagnetic radiation such as light and x-rays are directly emitted by electrons due to stimulation. The problem with this theory is that some radio waves, like those due to a 60 Hz ac current are more than 5 km in length. The question is how can the oscillation of an electron with a size of 2.82 x 10^{-15} m (classical radius of electron) result in a wave which is 5 x 10^6 m in length? This means that the 5 x 10^6 m wavelength is approx. 2.08 x 10^{23} times bigger than the electron or almost a quadrillion times the size of the electron, so it makes sense that such huge wave-lengths cannot be emitted directly by the electron. A very interesting fact about this huge 5 x 10^6 m wave length is that it possesses identical properties with that of a 'normal' photon, it travels at the speed of light, is electrically neutral, retains its energy etc., how can two such identical phenomena (i.e. high energy light waves and low energy radio waves) be attributed to two different causative factors? Namely that, high energy photons (i.e., optical photons) are directly emitted by excited electrons, while the formation and emission of low energy photons (like radio waves) is attributed to the oscillation of atoms in the conductor. Surely this is bad science? Photons of an energy higher than photons in the visible spectrum of light, such as gamma rays are the result of the catastrophic destruction of the atom or changes in the nucleus, or as in the case of x-rays from the sudden acceleration and braking of the electron. The Gestalt Aether theory states that there is a dimensional limit to the longest wave-length photon that an electron can emit and that the size of this wave-length is about 1.2×10^{-6} m which is considerably below the longest wave-length of visible light. Since electrons are known to emit and absorb photons with a wave-length of 7.17×10^{-7} m which is just within the limits of light that the human eye can discern, a wavelength for the conduction photon of 1.25×10^{-6} m as the longest wave-length that an electron can emit seems reasonable.

Photons in parallel and in series

So how do large wave lengths, such as radio waves form? The Gestalt Aether theory answer to this is that all electromagnetic waves greater than 1.2×10^{-6} m in wave length are composite waves, i.e. they are made up of joined or connected photons. This joining up of photons is made possible because of their dipole structure and can take place in two orientations, in series:

+   –

Photons connected in series

And in parallel

Photons connected in parallel

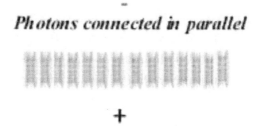

+

Photons connected in parallel

The largest photon wave length which an electron can emit is also, according to the Gestalt Aether theory, the photon which carries electrical energy and hence has been named the **"conduction "** photon. The conduction photon has a wavelength of 1.25×10^{-6} m Light when it travels through substances such as glass, crystal or water uses a method called "photon conduction" to pass through the substance, photons being rapidly absorbed and emitted during its passage. Yet when it comes to the conduction of electrical energy it is believed that it is electrons that serve to conduct electrical energy. Thus although every other form of energy related to electrons such as radiation and heat is mediated by photons, an exception is at present made in the case of electricity. Photons according to quantum mechanics cannot exist within a conductor due to the provisions of the Laws of the conservation of energy and momentum. Yet there are many facts which militate against this theory, electrons in a conductor have a drift velocity of 10^{-3} m one thousandth of a millimeter per second, while the current is seen to be established at approx. the speed of light or

3×10^{10} m/s (thirty billion millimeters a second), this is a huge discrepancy in speeds, how is it possible to rationalize this inconsistency?

The Domino theory of electrical conduction

The domino theory of electrical conduction does make such an attempt, it states that although the speed of individual electrons in a conductor under a difference of potential might be small, the density of electrons per centimetre remains unchanged, this means that a small change in position at one end of the wire, will immediately be reflected at the other end of the wire, thereby accounting for how fast electrical energy is conveyed in a wire.

Strictly speaking although this answer superficially seems to answer the question of how electrical energy is conveyed so fast in an electrical conductor, it does not fit all of the criteria, hence the adoption of the Bloch Band theory and Bloch wave functions. For instance, the domino theory does not explain how, in a wire carrying an alternating current, the electrons appear frozen in place; they do not appear to move at all. Since classical physics does not explain how electrical energy is conveyed, Gestalt Aether Theory turns to the quantum mechanics adoption of the Heisenberg Uncertainty Principle for an explanation.

Even though a simple inability to cope with conservation of momentum laws forbids free electrons from either absorbing or emitting electrons, the introduction of HUP, Heisenberg's Uncertainty Principle as it applies to energy and time might provide a solution. Heisenberg's Uncertainty Principle states that if an interaction takes place fast enough it can elude the conservation laws!

$$\Delta E \Delta t \geq \hbar.$$

What does this mean? It means simply that if a free electron travelling in a conductor emits a photon and then is able to re-absorb that photon fast enough, then it has avoided the conservation of momentum laws. In other words the energy multiplied by the time of the emission absorption process should be greater than or equal to h this gives an approximate time of 10^{-15} s for the emission and absorption process; a photon should be emitted and re-absorbed within a time interval of 10^{-15} s in order to avoid the laws of conservation of energy and momentum. A simple calculation shows that this time period provides no problem for electrons in a wire carrying a current.

$\Delta t \approx h / \Delta E$

$(6.62 \times 10^{-34}) / (1.6 \times 10^{-19}) = 4.13 \times 10^{-15}$ s

The Gestalt Aether theory states that electricity is conducted not by electrons but by photons. It is photons, not electrons are the fundamental charge carriers in a current carrying wire.

It should be understood that electrons in a wire carrying conductor at room temperature are extremely chaotic, electrons are shooting about in all directions at speeds in excess of 10^6 m/s although the average speed of electrons is zero due to collision with other electrons. When a difference of potential is applied the electrons are still moving around at 10^6 m/s and continue to collide with other electrons, however, now their average velocity is no longer zero but because of the applied difference of potential is now equal to what is known as the drift velocity or an average speed of approx. 10^{-3} m/s which is very slow but all electrons are now moving in a uniform direction towards the positive pole.

What happens to these photons, they have been emitted by the electron and have to be reabsorbed as soon as possible? As has been stated conditions in a conductor carrying a current at room temperature are chaotic, so much so that it is not possible for an ordered exchange of photons to take place, instead an electron which can be known as E1 emits a photon at time t1, it has to reabsorb a photon of the exact same value as the photon it had emitted before time t2 is reached. Time t2 as had been calculated is about 10^{-15} s, since electron E1 is under a difference of potential (both in AC and in DC circuits) it is not possible for electron E1 to always reabsorb, the same photon that it had emitted, so it absorbs a photon emitted by another electron we can call E2. The emitted photons are in a similar situation, they have to be reabsorbed by an electron requiring the exact energy which the emitted photons possess, in most cases, because of the difference in potential, it is not possible to immediately locate such an electron. In these circumstances such photons, exit the conductor and travel back in a loop to re-enter the conductor and find a receptive electron. When an emitted conduction photon leaves the conductor and attempts to loop back into the conductor, the virtual photons of the 'virtual photon aether' form into lines along the emitted conduction photons path. Each conduction photon has its own line of aligned virtual photons. Therefore, due to considerations of availability of space, lines of virtual conduction photons begin to form up in concentric layers outside the electrical conductor thereby forming the pattern of the lines of force we are familiar with. One must remember that a one centimeter cubed section of copper contains 8.5×10^{28} electrons. Therefore, even a small flow of current involves a huge number of electrons resulting in innumerable lines of force appearing around the conductor. So great is the number of interactions involved that these lines of force begin to form up around the conductor and stretch to a considerable distance into space. Each such line of force in the near field possesses the energy of one conduction photon which is 1.6×10^{-19} J. Therefore, each line of force in the near field possesses an energy of 1.6 x 10^{-19} J .

Photons are being emitted and then immediately being re-absorbed. Why? Because these photons can only be reabsorbed by electrons needing a similar quantum of energy to that which they carry, the nearest source of such electrons is within the conductor, so the photons which leave the conductor, re-enter the conductor forming what we know of as lines of force. Each line of force has the energy of one 'conduction' photon. Now what happens if the flow of the current suddenly stops; the photons that are able to enter the conductor and be absorbed do so, those photons unable to re-enter the conductor in time realign themselves in parallel and leave the conductor at the speed of light, in this case each line of force consisting of numerous conduction photons connected in parallel, shares the energy of a single 'conduction' photon. This orientation of the photons explains the energy difference between near and far fields.

This orientation of the photons (in series and in parallel) explains the energy difference between near and far fields. The **far fields** when they disconnect from the conductor and join up in parallel are **radio waves.** According to Gestalt Aether Theory, the fundamental **charge carrier** involved in the transmission of electricity **is not** as has erroneously believed for the past three hundred years the **electron**, but the **photon**. The Gestalt Aether Theory hypotheses is that photons are emitted and immediately re-absorbed by free electrons within the conductor (brouillion zone) in keeping with Heisenberg's Uncertainty Principle $\Delta e \Delta t \geq \hbar$ and are therefore able to avoid the Laws of conservation of energy and momentum which otherwise would prevent such emission and absorption of photons by free electrons present in the conductor:

Lines of force

One of the outcomes of this extremely rapid emission and absorption of photons by free electrons within the conductor are the lines of force that appear around any current carrying conductor.

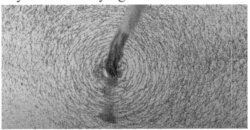

Electromagnetic lines of force around a wire carrying a current.

These lines of force are the manifestation of the 'virtual' photon aether which line themselves up in the direction of propagation of the real photon forming the complicated patterns that surround a conductor carrying electricity. Thus for those who constantly complain, where is this 'aether' that you are talking about? The answer is that it is literally all around you, surrounding the power line which comes to the house, in the electricity that runs the TV and the computer in and out of the numerous light fittings, in and around the compressor in the fridge and so on.

Radio waves

The photons emitted by free electrons in a wire carrying an electric current are 'relatively' low energy photons and have been named 'conduction' photons as opposed to optical photons which are normally emitted and absorbed by bound electrons in the atom and generally have larger energies. The following are the properties of a conduction photon:

The Quantum energy of the conduction photon $C_e^p = 1.6$ x 10^{-19} J

The wavelength of the conduction photon $C_\lambda^p = 1.25$ x 10^{-6} m

The frequency of the conduction photon Hz. $C_\omega^p = 2.4$ x 10^{14} Hz

Where C^p represents the conduction photon. C_e^p stands for the energy of the conduction photon , C_λ^P

156

represents the conduction photon wave length and C_ω^p is the conduction photon frequency. These are fixed values.

The energy of the Composite wave can be calculated by determining the number of conduction photons C_n^o in the composite wave and then dividing the energy of a conduction photon by the number of conduction photons present in the composite wave. Therefore if the wave-length of the composite wave is 0.75 m we can determine that it is made up of $C_\lambda^o / C_\lambda^p = C_n^o$
Therefore

$0.75 / (1.25 \times 10^{-7}) = 6.02 \times 10^5$ conduction photons

Therefore energy of the composite wave C_e^o is equal to the energy of conduction photon divided by the number of conduction photons present in the composite wave: $C_e^p / C_n^o = C_e^0$

$= (1.6 \times 10^{-19})/(6 \times 10^5) = 2.66 \times 10^{-25}$ J $=$ energy of composite wave (radio wave).

Using the equation $h\omega$ or hC_ω^o the same result can be found. For instance if $C_\omega^o = 4 \times 10^8$ Hz
Then $(6.62 \times 10^{-34}) \times (4 \times 10^8) = 2.65 \times 10^{-25}$ J $=$ energy of radio wave.

Therefore, it can be seen that the concept that radio-waves are composite waves consisting of numerous conduction photons connected in parallel is validated.

Near and Far fields

Now it is possible to see how the energy of the **far field** and the **near field** are produced. In the near field the 'conduction' photons are connected in series and each line of force holds the energy of one 'conduction' photon, so that in effect each line of force has an energy of 1.6×10^{-19} J. This fits in well with observed data and conforms with the flow of an electric current. Note that here the drift velocity of the electrons does not matter the lines of force in the near field each delivers the energy of one 'conduction' photon or 1.6×10^{-19} J.

Coming to the far field, here also the results are in line with observed data. In the far field the 'conduction' photons are connected in parallel thus each line of force in the far field consisting of numerous conduction photons connected in parallel, contains the energy of one conduction photon divided by the number of photons in the composite wave length. For example, given that we have a 0.9 m wave length in the far field then its energy will equal the energy of a conduction photon divided by the number of conduction photons present in the composite wave : $C_e^p / C_n^o = C_e^0$. The number of conduction photons present in the composite wave can be determined by dividing the composite wave length by the conduction photon wave length: $C_\lambda^o / C_\lambda^p = C_n^o$

Therefore $(0.9)/(1.25 \times 10^{-6}) = 7.2 \times 10^5$ conduction photons and the energy of the composite wave $C_e^o = C_e^p/C_n^o = (1.6 \times 10^{-19})/(7.2 \times 10^5) = 2.22 \times 10^{-25}$ J Energy of composite wave (radio wave). Similarly using $h\omega$ or $hC_\omega^o = (6.62 \times 10^{\wedge}-34) \times (3.333 \times 10^{\wedge}8) = 2.2 \times 10^{-25}$ J.

We can also see that if a voltage of one 1 volt is applied to the conductor and a current of 1 coulomb is made to flow then total current $I = 6.2 \times 10^{18}$ C (conduction photons) or 1.6×10^{-19} J x 6.2×10^{18} C = 0.999 Watts . Delivered at the speed of light C!

It is now possible to see that the longest wavelength photon that an electron can emit is the conduction photon C^p having a wave length of $C_\lambda^p = 1.25 \times 10^{-6}$ m. Radio waves are therefore a product of the far field formed by an alternating current and comprised of conductions photons that have not been able to resolve their energy by being absorbed by an electron within the conductor and instead link up with other similar photons in a parallel formation and exit the area at the speed of light. It can also be seen that while radio waves (composite waves) have a wave length based approximately on spatial dimensions, optical photons that are emitted directly by bound electrons within the atom have wave lengths that are determined temporally.

This theory of the manner in which electricity flows through a conductor, explains every possible aspect of electricity, without recourse to electric and magnetic fields. Including an explanation of why micro-waves and above cannot be absorbed directly by electrons. Yes sad but true, Maxwell's self-sustaining oscillating electric and magnetic fields are no longer needed, all that are needed is the aether and Faradays lines of force. This theory allows for the extremely accurate determination of current flow and for designing and making of electrical circuits that function with maximum efficiency. Most important of all it explains both how radio waves are generated and why these massively long wave-lengths possess the same properties as the photon.

The Standard Model explanation of how a current flows in a wire by means of being carried from electron to electron or electron to lattice by means of a wave function, is sadly lacking in both its logic and in its conclusions.

Static electricity and Dynamic electricity

The phenomenon of Static Electricity has been known since ancient times and was the subject of discussion by the Greek philosopher Thales as far back as 600 BC and was of keen interest to eighteenth and nineteenth century scientists such as Francis Bacon, Benjamin Franklin and Faraday. Dynamic electricity, involving the flow of an electric current through a wire on the other hand has only been known for the past one hundred and twenty years or so.

The concept of the electric field was first introduced by 19th century physicist Michael Faraday. It was Faraday's perception that the pattern of lines characterizing the electric field represents an invisible reality. The main difference between static electricity and dynamic electricity is that static electricity describes how separation of electric charges results in the formation of an electric field.

An **electric field** is the force that fills the space around every electric charge or group of charges. Electric fields are caused by **electrical forces**. Electrical forces are similar to gravitational forces in that they act between things that are not in contact with each other. Separation of electric charge results in different types of electric fields, since a charged object may be physically separated it is possible to study positive and negatively charged objects and the fields they generate:

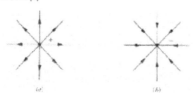

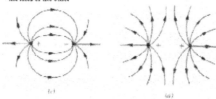

A positive charge will have the electric field radiating outwards from it (arrows pointing outwards) and a negative charge will have the electric field moving in towards it (arrows pointing in towards negative charge) like charges repel and opposite charges attract. So far it is obvious that there are a lot of similarities between electric fields and magnetic fields.

The study of electric fields is important to Gestalt Aether Theory which states that the lines of force represent the alignment of the 'virtual photons' of the Aether. If this is so, how is the presence of an electric field explained when no photons are being emitted and absorbed to account for the presence of the lines of force? The answer is that an electric field is only present when a difference of potential is present. When a difference of potential is established across two points separated by a dielectric a region of stress is created in the dielectric between the two oppositely charged points. The 'virtual photons' of the Aether line up in this region of stress within the dielectric giving rise to the 'electric' field.

The difference between an electric field and an electromagnetic field according to Gestalt Aether Theory is that the electric field is polarized resulting in alignment of the lines of force, while the electromagnetic field has energy flowing through the lines of force.

Standard Model Theory for the propagation of light.

Having seen a fairly comprehensive account of the Gestalt Aether Theory version of the propagation of light, an attempt will now be made to describe the Standard Model theory for the propagation of light, which is what the current theory for the propagation of light is.

Quantum mechanics as the evolving leader in the discipline of physics, needed to find a way of expressing electromagnetic radiation in terms of Max Planck's discovery of quanta; namely that EMR was made up of infinitesimal, discrete packets of energy. Instead of trying to find a new expression for EMR, quantum mechanics decided to adopt Maxwell's equations whole sale and to impose upon his theory of the propagation of electromagnetic waves, a quantum interpretation. A quantum interpretation involves not waves but particles, hence the quantum mechanical endeavour was to try to impose particle characteristics onto what was essentially a wave theory. More precisely, Maxwell stated that light is a propagating wave of self-sustaining, electric and magnetic fields. As one field waxed the other waned and vice-versa. The theory describes the interaction between the electric field and magnetic field. The direction of both the fields is perpendicular to each other. Maxwell also concluded that the electromagnetic wave travels at the speed of light. Trying to impose particle characteristics onto a wave system is a gigantic and unrewarding task, because the two systems have nothing in common, they are characterised by their differences not by their similarities. Converting Maxwell's equations to a form suitable for quantum mechanics involves a series of processes beginning with first quantization, followed by second quantisation, followed by normalisation and re-normalisation. What do all these processes mean and what is the end result? Here's a brief overview.

1) *First Quantisation:* refers to the standard approach of quantizing individual particles, in this case photons. In this framework, particles are described by wave-functions that evolve according to the Schrodinger equation. First quantization deals with the wave-particle duality and the behaviour of individual quantum

systems. (note) Schrodinger's wave equation works well for single particle situations, however with each additional particle that is needed, three additional spatial dimensions are required. This is a well-documented fact, freely admitted by both Max Born and Heisenberg. Though recent proponents argue that it is not dimensions that are involved it is multiple degrees of freedoms. This is an evasion rather than an attempt to address the problem. Even a cursory examination of the philosophical implications reveal that it is dimensions that are involved (i.e., actual spatial dimension) and not degrees of freedom.

An account of first quantisation can be found in: Griffiths, David J. *Introduction to Quantum Mechanics.* Pearson, 2016.

2) *Second Quantisation:* is a different approach used in quantum field theory (QFT) to quantize fields rather than individual particles. Instead of describing particles as discrete entities, second quantization treats particles as excitations of quantum fields that permeate space. This framework allows for the creation and annihilation of particles and provides a more natural description for many-particle systems present in EMR. Quantization of the Electromagnetic Field, involves imposing commutation relations on the creation and annihilation operators to ensure that the resulting quantum field theory is consistent with the principles of

quantum mechanics. This principle is also the present explanation for the propagation of radio waves. An alternating electrical signal in the antenna results in the spontaneous birth of a photon in the vicinity of the antenna, the photon undergoes spontaneous self-annihilation resulting in the creation of an electron and a positron, these two particles, since they represent matter and anti-matter undergo mutual self-annihilation almost immediately resulting in the creation of a photon of identical value as the original photon that underwent self-annihilation. Therefore, this is the Standard Model explanation of the creation and propagation of radio waves. As had been mentioned earlier the theory is well supported by mathematics involving creation an annihilation operators.

In second quantization, the electric and magnetic fields are represented by creation and annihilation operators. These operators create or destroy particles (quanta) associated with the electromagnetic field. During second quantisation the electromagnetic field is expanded in terms of its Fourier modes, which describe the field in terms of different wavelengths and momenta. (note) Each creation and annihilation process involves 1024 MeV approx., since there are an almost infinite number of photons all; undergoing these processes at the same time, it is a staggering amount of energy.

An account of Second quantisation can be found in:
Peskin, Michael E., and Schroeder, Daniel V. An Introduction to Quantum Field Theory. Westview Press, 1995.
And

Zee, Anthony. Quantum Field Theory in a Nutshell. Princeton University Press, 2010.

3) *Normalisation:* Could be described as a process, used in quantum mechanics to ensure that wave functions (or states) are properly normalized. Normalization ensures that the integral of the square of the wave function over all space (or some appropriate region) equals 1. This condition guarantees that the probability of finding a particle in any allowed region of space is unity. Normalization is achieved by dividing the wave function by a normalization constant, which is determined by integrating the square of the wave function and then taking the square root of the result.

Griffiths, David J. Introduction to Quantum Mechanics. Pearson, 2016.

This textbook provides a clear introduction to the principles of quantum mechanics, including the normalization of wave functions.

4) *Re-normalisation:* is a technique used in quantum field theory to handle infinities that arise in certain calculations. In quantum field theory,

interactions between particles can lead to divergent quantities, such as infinite values for certain physical quantities. Renormalization involves redefining these quantities in terms of experimentally measurable quantities, effectively absorbing the infinities into the parameters of the theory. Renormalization ensures that physical predictions from the theory remain finite and meaningful. It's a crucial aspect of quantum field theory, particularly in dealing with theories like QED where infinities arise in perturbative calculations.

Peskin, Michael E., and Schroeder, Daniel V. An Introduction to Quantum Field Theory. Westview Press, 1995. This book provides a detailed introduction to renormalization techniques and their applications in QFT.
And
Zee, Anthony. Quantum Field Theory in a Nutshell. Princeton University Press, 2010. - A comprehensive resource on QFT, including an accessible explanation of renormalization.

As can be understood from these quotes and citations of the Standard Model given above, quantum mechanics is hugely invested in protecting its interests in wave-particle duality, the Schrodinger wave equation and the wave-functions derived from it, and will go to any lengths to protect and safe guard, these insupportable and illogical esoteric ideas. For this reason, the Standard Model of quantum mechanics, now resembles a highly organised and strictly implemented religious rite rather than a scientific discipline. It is time for a change 'out with the old, in with the new.' Do you agree? (Note: The Term Standard Model is a hold-all name for quantum mechanics, Quantum Electrodynamics and Quantum field theory).

One has to question the degree of entitlement and isolation from reality that is needed to make statements such as an object can be in two places at once (super position) or that two objects that are spatially separated (i.e. one object at one side of the Universe and the other object at the other side of the Universe) can instantaneously influence each other's behaviour.

Consequences and Implications of Gestalt Aether Theory or GAT

The Gestalt Aether Theory of electrical conduction and light propagation has several positive points in its favour. To begin with energy is conserved. The theory allows for stable energy levels of electrons at an energy of 1.6×10^{-19} J, for the conservation of energy within the electrical conductor. The same consideration applies to bound electrons within the atom. The modality implemented in the constant absorption and emission process further allows the electron to dynamically regulate its energy and again the same applies to bound electrons within the atom. This characteristic of having a universal application for both bound and free electrons is what lends the theory strength.

Emitted photons have a fixed energy corresponding to the energy absorbed by the electron. This emphasises the idea that photons have discrete and quantised energy. Further the emission spectrum of an atom would be characterised by specific wave-lengths corresponding to the energy transition defined by Gestalt Aether Theory. It explains both how light travels in a rectilinear manner as well as how it spreads out according to the inverse square law. By stating that the electron's motion towards and away from the nucleus follows a classical trajectory, influenced by the laws of reflection and momentum conservation, and simultaneously explains the mechanism by which the electron is able to maintain a very high rate of photon absorption and emission. Most importantly this theory allows for the one on one interaction of electron and photon.

Similarly the orbit of the electron around the nucleus is revived and follows the same principle originally stated by Neils Bohr, namely that the orbit of an electron depends on classical factors such as its mass, momentum and energy and is in keeping with observed emission and absorption spectra of atoms. At the same time the Pauli Exclusion Principle also holds good for distribution of electrons through the atom. This theory is available for the whole range of elements, from the simple Hydrogen atom to the most complicated atoms such as lead and uranium.

The theory also gives a definitive explanation on the nature of radio-waves, their formation and propagation that is well supported and sustained through observation and between the electron and the nucleus. The electron does not collapse into the nucleus despite being propelled towards it, due to the neutralization and rebound mechanism. This is similar to the situation when an electron is propelled towards the anode, it is not repelled at the last minute. Because a difference of potential has to exist between the cathode and electron it is not possible in this instance to determine what the situation would be if a negatively charged particle and a positive charged particle both with the exact same charge meet, in all probability the two charges would be neutralised and there would be no effect at all. In the case of a negatively charged electron approaching the positively charged nucleus, the two charges would temporarily be neutralised.

The novel idea introduced in the theory is the manner in which the electron can self-regulate its energy through the constant emission and absorption of virtual particles in a process of self-interaction and through virtual interactions with other electrons, in multi-electron atoms. The theory explains why the electron does not radiate away its energy and spiral into the nucleus as was originally conjectured by Larmor. Most importantly the theory opens the way for the setting aside of wave-particle duality, together with its dependent theories of superposition, multiple dimensions, wave-functions etc, and has a cleansing effect on physics, and does away with much of the esoteric content that has dominated physics for the past 100 years.

Further the introduction of the concept of 'conduction' photons as the charge carriers in an electric current limits the size of the largest wave-length photon (although the physical dimensions of the photon remain the same as that of any other photon)and brings both radio-waves and light under the same paradigm.

The consequences of this theory on photon emission and absorption have significant implications for our understanding of atomic behaviour, energy conservation, and electromagnetic radiation. By offering a classical perspective that complements quantum mechanics, Gestalt Aether theory provides a framework for exploring the dynamics of electron-photon interactions and their impact on atomic stability and radiation characteristics.

The Gestalt Aether theory, possesses great potential in leading to new insights and advancements in atomic physics. The theory provides a detailed classical mechanism for photon emission and absorption that results in highly directional and individual coherent photon streams. However, when looked at overall, with perhaps 2×10^{23} atoms all emitting in different direction and at different angles the light is anything but coherent. This is in keeping with observed behaviour of light. At the same time because these frequencies are so close together it is possible for different frequencies to temporarily interconnect to give rise

The superposition of two closely related frequencies of light to give a new frequency.

to light of a new frequency, in the same way that two musical notes can combine to give a new note. The two frequencies can also separate into their original frequencies with equal ease.

The Gestalt Aether Theory, explains every aspect of light from its inception in electrons within the atom to the manner in which photons propagate as waves. The Gestalt Aether Theory (GAT) explains how electrons and photons are able to interact on a one on one basis. The photon is emitted by a radiating source and is directly absorbed by a receptive electron in an atom. In order for an electron to be receptive to a particular photon, might involve nothing more complicated than that electron being at the right energy level within the atom to absorb the photon. The advantage of the GAT, is that it explains all aspects of light without once venturing into thorny subjects such as wave particle duality or making the supposition that at the level of the very, very small things behave differently from their behaviour in the macro world. There exist simply no grounds for such a premise to be made.

Chapter 5 : The Cosmic Microwave Background Radiation and Dark Matter

George Gamow and the Big Bang Theory

In the mid 1940's George Gamow, along with his students Ralph Alpher and Robert Herman, worked on developing the Big Bang theory, which had gained prominence during the 40's and 50's. The Big Bang theory was put forward by Belgian priest, scientist and Nobel prize winner Georges Le Maitre (1894 – 1966) and was based on Einstein's relativity and the concept of an expanding Universe. The Big Bang theory posits that the universe began as an extremely hot and dense singularity around 13.8 billion years ago. In this initial state, the universe was filled with a hot, dense plasma of particles, including quarks, gluons, electrons, and photons (light).

Georges Le Maitre inventor of the Big Bang Theory

Gamow and his team were particularly interested in understanding the conditions of the early universe and the formation of elements, a field known as nucleosynthesis.

In 1948, Alpher and Herman predicted that the early universe would have been extremely hot and dense, leading to a thermal radiation background as it cooled. This radiation would have been in thermal equilibrium with matter in the early stages of the universe. They calculated that the residual radiation from this early hot phase would have cooled down as the universe expanded, eventually settling into the microwave region of the electromagnetic spectrum. They estimated the temperature of this background radiation to be around 5 K, which corresponds to wavelengths in the microwave range.

Throughout the 1950s, the idea of a residual thermal radiation background remained largely theoretical. There was no concerted effort to detect it experimentally, partly because the necessary technology was not yet fully developed and partly because the scientific community had not universally accepted the Big Bang theory. Advances in microwave detection technology began to make it feasible to search for the predicted background radiation.

1964 Discovery of CMBR by Penzias and Wilson:

Arno Penzias and Robert Wilson, working at Bell Labs, were conducting experiments with a horn antenna designed for satellite communication. They encountered persistent noise in their data, which they could not eliminate. After consulting with scientists, including Robert Dicke and Jim Peebles at Princeton University, who were also searching for the CMBR, Penzias and Wilson realized that their unwanted noise matched the predicted radiation. They measured this radiation to have a temperature of approximately 3.5 K (later refined to 2.725 K), consistent with the predictions made by Gamow, Alpher, and Herman.

The wavelength of 1 mm that was predicted for the relic radiation from The Big Bang, corresponds to a frequency of about 300 GHz and a temperature of about 2.725 K, fitting well within the microwave region. This specific wavelength was significant because it falls in a range where the Earth's atmosphere is relatively transparent to microwave radiation, making it easier to detect from ground-based instruments. The choice of 1 mm as a core signal was not initially a deliberate target by Gamow's team but rather a result of the theoretical predictions and subsequent confirmation by observational data.

The prediction and eventual discovery of the CMBR were pivotal moments in cosmology. Gamow and his team's theoretical work laid the groundwork, while Penzias and Wilson's serendipitous discovery provided the experimental evidence needed to confirm the Big Bang theory, establishing the CMBR as a cornerstone of modern cosmology.

Is it possible the CMBR theory is wrong?

Atoms are made up of 99.99999999 % empty space. How is it that it is not possible to just walk straight through walls considering that we are made up of 99.999999999 % empty space? We live in a world of electrostatic attraction and repulsion. When an attempt is made to try to pass through a wall, we first experience a neutral force where the electron- proton attractive force is neutral because both forces are in equilibrium, as the wall gets closer the electron-electron repulsion force is stronger than the electron-proton attractive force because the electrons are much closer to each other, and it is not possible to advance further.

One aspect of living on earth that is often ignored is just how volatile our environment is. If we pick up a pencil, trillions of electrons have to re-adjust themselves in order to maintain equilibrium. Just the act of walking requires amazing feats of balance featuring an untold number of electrostatic interactions. In fact, every single movement or action that is made results in the displacement and re-arrangement of the electrostatic forces that surround us, this involves trillions of electrons. Given that this is so, it is strange that the theories we have of the Universe reflect nothing at all of this aspect of existence. The theories of the Universe that are at present in wide-spread use view the Universe as a place that is completely at rest. This is the only explanation that exists for the present wide-spread acceptance of the Cosmic Microwave Background Radiation (CMBR). This theory seems to take for granted that there is no activity at all in the rest of the Universe and that Earth is the only place in the whole of the wide Universe, where the act of picking up a pencil displaces trillions upon trillions of electrons. Surely such disturbances exist in the rest of the Universe, more particularly in the unimaginably massive clouds of hydrogen that surround us in every direction, unquestionably currents will be set up, resulting in radio transmissions in many different frequencies. The whole of the Universe was formed out of such disturbances in the massive hydrogen clouds. As to the 2.7 K constant temperature that is held up to show that the CMBR is a kind of Black Body Radiation, surely it can be taken for granted that the outer reaches of the Universe are cold.

One of the founding principles of science, is that light can never be at rest, it and all electromagnetic radiation are either moving at the speed of light or have been absorbed. This means that the CMBR which is relic radiation from the Big Bang has been travelling for billions of years. According to the theory the CMBR is coming from everywhere in the Universe? Is it possible? Perhaps the biggest factor militating against the 1mm signal that has been designated as the CMBR is the fact that it is possible to pick up these signals with any old radio equipment, the white static on old cathode ray tubes was also supposed to be from the same source: the CMBR. But seriously this should not be the case at all. Radio astronomers use massive, very sensitive antennas to pick up radio signals emanating from stars and galaxies that are just a few hundreds of light years distant and then the signal is just the tiniest whisper. To pick up signals that were formed at a time just 300, 000 years after the Big Bang as is supposed to have happened with the Big Bang, should need the most sophisticated equipment that we possess, akin to the deep space network. The fact that such sophisticated equipment is not necessary and that it is possible to be able to receive relatively loud signals without the benefit of sophisticated equipment is a big indicator that the so called CMBR is not relic radiation from the Big Bang at all but that it is very much a symptom of the present. In other words what has been designated as being the CMBR, is actually not the CMBR. Another very compelling fact militating against the CMBR being relic radiation from the Big Bang can be found by using the Doppler shift. The velocity of an object can be determined by the amount the spectrum is shifted. The amount of the shift is the source's velocity relative to the observer. This major clue tells astronomers if an object is moving towards us or away from us, and at what speed. In the case of the CMBR this shift can be calculated as follows:

$V_{source} = c (1- (\omega_o/\omega_{obs})$ If we take the original frequency ω_o to be 500 THz and the observed frequency to be 3×10^{-11} m . Then it is possible to see that:

$(3.14) \times (1- (5 \times 10^{14})/(10^{-11})) = -499700000000$ m/s This means that the CMBR source is moving away from us at the rate of :

$(4997 \times 10^8)/(3 \times 10^8) = 1667$ times the speed of light! Or in other words the CMBR represents a signal that is receding away from us at a speed that is 1667 times the speed of light approximately. This result makes it extremely, unlikely that the CMBR is from the time period that is designated.

The CMBR is not the thermalised (static?) radiation that everyone seems to accept without question. Instead, it is a reflection of present events that are taking place in the Universe, it correctly reflects the distribution of matter in The Universe. Think of the huge, massive beyond belief, clouds of gas in which the CMBR has its origins. Is it possible to think that these massive clouds of gas are static, that no movement at all takes place within them? No. You can be sure that currents are constantly being generated within these massive clouds of hydrogen that give rise to radiation and constant interactions with light and electromagnetic radiation coming in from stars and galaxies taking place. If this were not the case, there would be no evolution: stars and Galaxies would not evolve from these hydrogen clouds.

Why is the CMBR so homogenous? In order to understand why the CMBR is homogenous one must have some comprehension of the size of these clouds of hydrogen which are more or less homogenous in composition. This being so, if the so called CMBR emerging from these clouds were not homogenous, it would be cause for concern and investigation.

What is the CMBR?

No-one, not even the most sceptical of persons, can deny the vastness of the Universe. Our own

Milky Way Galaxy, is itself 105,000 light years across approximately, give or take a few hundreds of light years. By far the biggest single component of the Universe are the vast clouds of hydrogen that populate the Universe. The size of these clouds of hydrogen beggars the imagination, a single cloud of hydrogen might contain hundreds of thousands of Galaxies of the size of our Milky way Galaxy.

It is then a matter of some concern that the so called CMBR is so widely accepted as being the defining signal from the Big Bang. I have no problem in accepting that such a signal exists, what I have a problem with is its designation as relic radiation left over from the Big Bang. This does not make sense. Why? For one thing, electromagnetic radiation does not stop moving ever, it is always moving at the speed of light, it never stops. To imagine that the CMBR (relic radiation or not) would just hang about in one place is not justifiable. If it is still travelling, where is it travelling to? For another, we would have to accept that the present day Universe is absolutely quiescent. Why? This is because the present day Universe is made of hydrogen and the signal of the CMBR falls squarely within the Hydrogen spectra. So what are we to believe, are we to take for granted that the present day Universe is absolutely quiescent and all of the signals in the hydrogen spectra being received on earth are due to the relic CMBR and nothing from the present day Universe? Surely, this is absolutely illogical? We know, that the present day Universe is not quiescent, we know that vast movement must take place within the clouds of hydrogen gas that result in every possible signal over the hydrogen spectra and even beyond. The fact is that the CMBR is not relic radiation at all and instead is very much a signal that is being produced by our present day Universe.

What substance then would point to the existence of the Big Bang and a finite Universe. I put it to you that the most significant indication of a Universe that had its birth in the Big Bang, would be Dark Matter, yes, with its very low energy of 10^{-51} J it would be absolutely permeable to all matter, by the same token applying Heisenberg's Uncertainty Principle as it applies to time and energy, it is possible to see that given their extremely low energy, the lifetimes of these particles becomes indefinitely long. Dark Matter, if it is made up of extremely low energy electric dipoles is a much better candidate for relic radiation from the Big Bang than is the CMBR.

Dark Matter

Dark matter is one of the most fascinating and elusive components of the cosmos, contributing to some of the most profound mysteries in modern astrophysics and cosmology. Although it is invisible and does not emit, absorb, or reflect light, dark matter constitutes about 85% of the total mass of the universe. Its presence is inferred from gravitational effects on visible matter, such as stars

184

and galaxies, indicating a substantial influence on the large-scale structure of the cosmos. Dark Matter has proved to be undetectable to all modern forms of investigation. It is invisible, it is odourless, it is non-tactile to the point of being non-existent. It is colourless, it has no mass, it is electrically neutral and cannot be detected by electric or magnetic fields. In other words Dark Matter is a mystery the existence of which, while being undeniable, is one of the biggest unsolved enigmas of the Universe today. What is it? What is it made of? Why does Dark Matter have such low interaction with matter? How can a substance that has no mass and cannot be weighed, exert such a huge influence as to be able to hold Galaxies together? An even more extraordinary property of Dark Matter, and as far as I am concerned its most remarkable attribute, is its ability to allow all kinds of Electromagnetic radiation, including light, radio-waves and even Gamma Rays and x-rays (which are an extreme form of electromagnetic radiation) to travel through it without offering the slightest opposition or interaction. We are able to receive light from stars that are hundreds, thousands and even billions of light years away, without the slightest interference taking place. This is truly remarkable when one takes note of the fact that Dark Matter accounts for 85% of matter in the Universe, as testified to by reliable Scientific authorities. Transmissions made on the surface of the earth are met with interference and opposition at every turn. Everything from the air to solid objects to other radio-waves and electromagnetic radiation interferes with the propagation of electromagnetic radiation. So for Dark Matter to enable the transmission of electromagnetic radiation of all frequencies across the vast breadth of the Universe, without offering any interference is truly remarkable.

Dark Matter in the solar system

If one looks dispassionately at the properties of Dark Matter enumerated in the preceding paragraph, one finds that the properties of Dark Matter are remarkably similar, I would go so far as to say identical, to the physical properties that were attributed to the aether: the medium through which light supposedly propagated on earth. By the term aether, I am not referring to the ludicrous concept of the Luminiferous aether, which was a concept of the late 19th Century, but to the earlier concept of an aether as a medium for the propagation of light. Indeed, the similarities between Dark Matter and the discarded concept of the aether, might be more than just a coincidence, since Dark Matter occupies 85% of the mass of the Universe, the odds are heavily in favour of our Solar System lying within that 85%, this being so it leads to the inevitable conclusion, that what we had taken to be the aether, was in fact, Dark Matter. Just as the aether had remained elusive and undetectable to every form of investigation, so to Dark Matter has remained undetectable even to all the extraordinary technical equipment available to us today.

What does this mean? It means, for instance that it is impossible to collect Dark Matter, put it on a weighing machine and attempt to weigh it. For one thing no vessel could contain Dark Matter, if you did succeed in scooping up some Dark Matter in a vessel it would immediately, travel through the material of the vessel, so that there would be nothing there to weigh. Say, in the future, maybe in an environment completely saturated with Dark Matter, it were possible to collect some Dark Matter and get it onto a weighing machine. It would do absolutely no good, Dark Matter would just sink right through the weighing machine. It would not be possible to touch Dark Matter to ascertain that you actually have hold of it, it is colourless and invisible. In effect undetectable.

The concept of dark matter first emerged in the 1930s through the pioneering work of Swiss astrophysicist Fritz Zwicky. While studying the Coma Cluster of galaxies, Zwicky observed that the galaxies were moving at such high velocities that they should have dispersed if they were held together only by the visible matter that was present. He postulated the existence of an unseen mass, which he termed "dunkle Materie" (dark matter), to account for this discrepancy. Zwicky's calculations suggested that the mass of dark matter within the cluster was far greater than the mass of visible matter, leading to the idea that a significant portion of the universe's matter is invisible.

In subsequent decades, the evidence for dark matter continued to accumulate. As late as the 1970's, American astronomer Vera Rubin's definitive work on the rotation curves of spiral galaxies provided further compelling evidence. Rubin found that the outer regions of galaxies were rotating much faster than could be explained by the visible matter alone. This anomalous rotation suggested that a vast amount of unseen matter was exerting a gravitational pull, keeping the galaxies from flying apart. Rubin's observations reinforced the notion that dark matter is a fundamental component of galactic structures. Vera Rubin's eventual estimate of the amount of Dark Matter in the Universe was close to 90%.

Despite the robust evidence for its existence, the exact nature of dark matter remains one of the biggest unanswered questions in science. Dark matter is hypothesized to be composed of non-luminous particles that interact primarily through gravity and possibly through weak nuclear forces. Various candidates have been proposed, including Weakly Interacting Massive Particles (WIMPs), axions, and sterile neutrinos, but none have been conclusively detected. Efforts to identify dark matter particles are ongoing, with numerous experiments conducted deep underground, in space, and using particle accelerators.

Dark matter plays a crucial role in the formation and evolution of cosmic structures. In the early universe, it acted as a scaffold for ordinary matter to clump together, leading to the formation of galaxies and galactic clusters. Simulations of the universe's evolution, which include dark matter, accurately reproduce the large-scale distribution of galaxies observed today. These simulations show that dark matter forms a vast cosmic web, with galaxies and galaxy clusters residing at the intersections of these filaments. However, this idea of the clumping together of Dark Matter into various areas and regions of the Universe is as ridiculous a concept as the Luminiferous aether proved to be. It can clearly be seen that in every direction, one can turn to on earth or even in space as demonstrated by our satellites and space craft, electromagnetic radiation experiences little to no interference. This single circumstance goes a considerable way towards proving that the distribution of Dark Matter throughout the Universe is uniform.

The study of dark matter is not limited to astrophysics and cosmology; it also intersects with particle physics and general relativity. Researchers are exploring the possibility that dark matter might interact with ordinary matter through forces beyond gravity, leading to potential breakthroughs in our understanding of fundamental physics. Additionally, modifications to Einstein's theory of general relativity, and MOND (Modified Newtonian Dynamics) theories dealing with modified gravity, have been proposed to explain dark matter's effects without invoking new particles.

In conclusion, dark matter remains one of the most intriguing and essential components of the universe. From its initial postulation by Fritz Zwicky to the groundbreaking observations by Vera Rubin, dark matter has profoundly influenced our understanding of the cosmos. While its exact nature continues to elude scientists, ongoing research across multiple disciplines promises to shed light on this mysterious substance. As we delve deeper into the nature of dark matter, we move closer to unraveling the fundamental workings of the universe itself.

Gestalt Aether Theory and Dark Matter

Yet that could be far from being the whole story, for 240,000 years before re-combination began, photons had been pumped out of matter at an extraordinary rate. The question is what happened to those photons ? At some point in time, apart from heat, the Universe must have also been flooded with light. The question is what happened to that light? Light travels at

$$3 \times 10^5 kms/sec,$$

while matter travels at thousands or tens of thousands of kilometres per hour, if light existed, and since by (the existing) definition a photon will travel for ever until it is absorbed by an electron in its path, shouldn't this light have outpaced all matter and advanced several billions of light years over the border of the Universe? At this point it is important to point out that in every sense that light or anything else could travel past the boundaries of the Universe is nonsensical since by definition there is nothing there. The Big Bang theory does not attempt to explain what initiated the creation of the universe, or what came before the Big Bang, or even what lies outside the universe. All of this is generally considered to be outside the remit of physics, and more the concern of philosophy. Given that time and space as we understand it began with the Big Bang, the phase "before the Big Bang" is as meaningless as "north of the North Pole".

The Second Law of Thermodynamics , on the other hand, lends theoretical (albeit inconclusive) support to the idea of a finite universe originating in a Big Bang type event. If disorder and entropy in the universe as a whole is constantly increasing until it reaches thermodynamic equilibrium, as the Law suggests, then it follows that the universe cannot have existed forever, otherwise it would have reached its equilibrium end state an infinite time ago, our Sun would have exhausted its fuel reserves and died long ago, and the constant cycle of death and rebirth of stars would have ground to a halt after an eternity of dissipation of energy, losses of material to black holes , etc

Historical perspectives of the Universe:

Until the time of the formulation of the Big Bang Theory, the assumption of a static universe had always been taken for granted by astronomers. To put things into perspective, for most of history it had been taken for granted that the static earth was the centre of the entire universe, as Aristotle and Ptolemy had described it many centuries ago. It was only in the mid-16th Century that Nicolaus Copernicus showed that we were not the centre of the universe at all (or even of the Solar System for that matter!). It was as late as the beginning of the 20th Century that Jacobus Kapteyn's observations first suggested that the Sun was at the centre of a spinning galaxy of stars making up the Milky Way. Then, in 1917, humanity suffered a further blow to its pride when Curtis Shapely, revealed that we were not even the center of the galaxy, merely part of some unremarkable suburb of the Milky Way (although it was still assumed that the Milky Way was all there was). Some years later, in 1925, the American astronomer Edwin Hubble stunned the scientific community by demonstrating that there was more to the universe than just our Milky Way galaxy and that there were in fact many separate islands of stars - thousands, perhaps millions of them, and many of them huge distances away from our own. These galaxies had been noted in the past under the name of constellations but no-one had ever suspected that they were galaxies on the scale of the milky-way. Suddenly the Universe became an infinitely large place.

From light to aether and back again:

Given the physical structure of the photon as being massless, electrically neutral, indivisible quanta of energy as described in the preceding Chapters of this book, and the huge disparity between matter to photon ratio, it was suggested that the massive amounts of photons that had been created at the time of the Big Bang, could not cross over the borders of the Universe and therefore accumulated within the Universe. It is these very low energy photons that constitute Dark Matter.

All of the matter in the Universe would represent just a tiny fraction of the light that filled the Universe, the chances of that light being absorbed by matter are therefore negligible. What happened to all this light, to this literally innumerable collection of photons? One possibility is that photons as they were created were pushed outwards spreading further and further out towards the edges of the Universe. Since the model of the photon according to the Gestalt Aether Theory presented in this book, is that they are di-pole and solenoid in form, it means that the wave like structure they possess allows them to join together, when they do this there is a natural dispersal of energy throughout the photon medium. The photon energy began to disperse throughout the joined photon structure, until finally it was left with such low energy (10^{-51} J per photon) that it for all purposes ceased to exist. These extremely low energy 'virtual photons' could theoretically exist forever. How is this possible? This possibility is due to one interpretation of the Heisenberg Uncertainty Principle that states that if the energy of the object is sufficiently small the time over which it can exist is indeterminate. Under these conditions the early photons could exist practically forever with a life span similar to that of the electron or proton. Thus the Universe was filled with these practically stationary di-pole electromagnetic points, that were oriented at random: in the presence of a real photon, these 'virtual' photons of the universal field (aether) line up in the direction of propagation of the real photon, forming a line whose ends rest on the shoulders of infinity and the energy of the real photon travels along this line. Note that it is the structure of the photon, is it wave or is it particle OR is it wave and particle simultaneously, that makes the formation of a Universal field (aether) that permeates every part of the Universe possible. The photons of the early Universe were undetectable, and still are undetectable today, because of their extremely low energies. It was only when the whole of space was completely permeated by the universal field (aether) that the propagation of light as we understand it today became possible. This is the universal virtual photon field (aether) that pervades the Universe.

One way in which these early photons (light) could continue to exist in the Universe is if it experienced a loss of energy, then in accordance with Heisenberg's Uncertainty Principle : $\Delta E \, \Delta T \geq h$, these very low energy ($10^{-51 \, J}$) virtual photons could exist as had been calculate elsewhere for an arbitrary long time on the other order 2.6 x 10^{10} years or something in the region of twenty billion years.

Gestalt Aether Theory holds that the virtual photon aether that it refers to is in actual fact Dark Matter. If one gives due consideration to this idea, it soon becomes apparent that this idea fits in well with the data that has already been collected on Dark Matter, it explains the very low interaction with matter, it accounts for the fact that Dark Matter appears to have no mass, it is undetectable. Yet at the same time, Dark Matter allows for the free passage of all types of electromagnetic radiation without offering any interference or resistance. It also exerts a gravitational force. How can a massless, electrically neutral particle of extremely low energy, exert a gravitational force?

The Virtual Photon Field: Unveiling Dark Matter's Role in the Missing Mass of the Universe:

In order to assess whether virtual photons could account for the missing mass and in effect give a name to

Dark Matter. It was decided to see if the possibility existed that Dark Matter was indeed composed of virtual photons. There are a few problems that have to be addressed. To begin with the mass of the Milky way galaxy comprises the mass of all the baryonic matter in the milky galaxy together with the mass of the Dark Matter in the milky way galaxy. Nevertheless, an attempt can be made, if calculate the volume of a cylinder as being = to $V = \pi r^2 h$ then the volume of a photon would be equal to approximately

$$3.14 \times ((10^{-16})^2) \times 10^{-6} = 3.14 \times 10^{-38} \, m^3$$

However, this does not give an accurate answer because we have to consider the area of influence of the photon, considering its very low energy of 10^{-51} J an area of influence amounting to 10^{-6} m^2 seems reasonable. Given that the volume of the milky way galaxy is 3.33 x 10^{60} Kg. it is possible to calculate the percentage of Dark Matter.

To assess whether virtual photons could contribute significantly to the missing mass in the Milky Way galaxy, the following calculations were carried out, based on given information:

1. Energy of a single virtual photon: $E = 10^{-51}$ J

2. Volume occupied by a single virtual photon: $V = 3.14 \times 10^{-18}$ m³ Note that the volume of the virtual photon is calculated on the area of direct influence of the virtual

196

photon which is in the region 10^{-6} m. If we take the volume of the milky way galaxy to be 3.33 x 10^{60} m^3 Then the number of photons in the milky way galaxy would be:

(3.33 x 10^{60}) / (3.14 x 10^{-18}) = 1.09 x 10^{78} photons in the milky way galaxy approx.

The mass of a photon with an energy of 10^{-51} J can be calculated using the equation

$E = mc^2$

So here 10^{-51} / c^2 = 10^{-51} / (9 x 10^{16}) = mass of virtual photon = 1.14 x 10^{-34} Kg

Given that the mass of the milky way galaxy is 3.32 x 10^{42} Kg and the number of photons each with an energy of 10^{-51} J needed to balance this is 3.32 x 10^{42}/ 1.112 x 10^{-34} = 2.9 x 10^{76} photons. This number is very close to the number of virtual photons calculated to be in the Milkyway galaxy; the number of virtual photons that had been calculated to be in the Milkyway galaxy being in the region of 1.09 x 10^{78} photons.

Therefore, even though the virtual photons as described by Gestalt Aether Theory have a very low energy they can still account for 85% - 90% of the mass in the milky way Galaxy.

Gestalt Aether Theory and the laws of thermodynamics:

In the context of a universe governed by a virtual photon aether, also known as dark matter, the laws of thermodynamics can be reinterpreted through the lens of

197

this fundamental medium. If we posit that the universe was created with a fixed amount of energy that cannot be created or destroyed, but only fluctuates within this fixed energy reservoir, it offers a novel perspective on how energy dynamics and thermodynamic principles operate.

The laws of thermodynamics, particularly the principles of energy conservation and entropy, can be viewed through this fixed energy lens. Thermodynamic equilibrium would be achieved not by creating or destroying energy but by redistributing it within the fixed reservoir. The interactions of matter with virtual photons would result in energy exchanges that are balanced within this constant energy framework. Entropy, which measures the degree of disorder or randomness in a system, could be understood as a reflection of the fluctuations within the fixed energy reservoir. As energy is redistributed among various forms, the system's entropy would adjust accordingly. In this model, entropy increases or decreases as a result of energy fluctuations rather than changes in the total energy of the universe.

The fixed energy concept implies that conservation laws are upheld on a cosmic scale. The apparent changes in energy are reconciled within the framework of the virtual photon aether, which ensures that the total energy remains constant. This reinterpretation aligns with the classical conservation of energy while accommodating the dynamic interactions facilitated by virtual photons. The presence of a fixed energy reservoir mediated by virtual photons could contribute to the stability and evolution of cosmic structures. The balance of energy fluctuations could influence processes such as galaxy formation, cosmic inflation, and dark matter interactions. By regulating energy distribution through virtual photons, the universe maintains a consistent thermodynamic environment over vast scales of space and time.

In a Universe characterized by a fixed amount of energy mediated by a virtual photon aether, the laws of thermodynamics are reinterpreted through energy fluctuations within a constant reservoir. Virtual photons play a crucial role in redistributing energy, maintaining equilibrium, and regulating entropy. This model provides a framework for understanding how thermodynamic principles apply in a universe where energy is neither created nor destroyed but continually redistributed within the fixed energy pool.

A new perspective of existing ideas can yield new ideas.

Chapter 6:
Olber's Paradox and the Hubble Shift

The wonder of the night sky

Everyone has looked up at the night sky at one time or another and marvelled at the stars, planets and galaxies that are revealed. In the cities and urban landscapes, the night sky seems remote and of little interest. On most nights the night sky is not even visible, because of pollution caused by incessant traffic but more importantly because street lighting is kept on till such late hours, in some cities even being left on for twenty-four hours a day that the night sky is totally obscured. The exception to this rule is probably for those people living in high rise buildings who are high enough up to see a relatively unobscured night sky. For people who have the time, and the opportunity, to look at the night sky, it is an incredible sight. Up in the mountains or out in the country side the night sky presents an incredible spectacle. Virtually every inch of the sky in whichever direction one chooses to look is filled with stars. Almost equally striking are the myriad colours that are on view, ranging from reddish and pinkish to stars to all shades of blue and dazzling white and a few that seem yellow.

For anyone who has the time to look at the stars and to wonder what they are seeing, the first point that strikes the imagination, are the sheer numbers of stars that are present. If it is a dark night and it is a new moon, the second thought that follows, is to wonder, why, if there are so many stars present, there is so little light, because when you look down there is hardly enough light to see the ground, everything is pitch black on such a night. Surely, with all the hundreds of millions of stars that are up there in the sky, the light that they create should be much brighter than that of the moon? What a pity that it isn't so. These thoughts are soon forgotten and one very quickly accepts that it is just the way things are. Yet, if one thinks about it many of those questions are relevant. Why isn't the sky lit up at night?

Olber's Paradox

Somewhat disappointingly, calculations show that in the normal view of a night sky only about 2500 stars are visible but if one is far out in the country away from bright lights it is possible to see thousands of stars, in fact literally uncountable numbers of stars are visible.

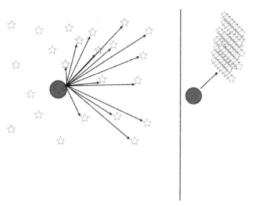

Olber's Paradox

When Olber formulated his paradox, theories of a finite Universe were unknown the Universe was thought to be infinite and eternal. The argument is as follows: In a Universe consisting of unbounded space populated endlessly with luminous stars, a line of sight in any direction from the eye, extended into the depths of space, eventually intercepts the surface of a star. Visible stars should therefore cover the entire night sky without any separating gaps. Hence, the riddle of cosmic darkness: Why is the sky dark at night?

"Basically the argument for why this is so is that " the succession of stars was endless and had existed for an infinite amount of time, therefore the background of the sky ought to present a uniform luminosity, like that displayed by the Milky Way Galaxy – since there would be absolutely no point, in all that background of the night sky, at which a star would not exist. The only mode, therefore, in which, under such a state of affairs, we could comprehend the voids which our telescopes find in innumerable directions, would be by supposing the distance of the invisible background so immense that no ray from it has yet been able to reach us at all." Attributed to Lord Kelvin

Thus the argument that was used to solve Olber's paradox of the dark night sky is that the Universe is so immense that the light from a large number of stars has not yet reached us. Yet Lord Kelvin's statement stating that numerous voids are present in every direction when the night sky is viewed through a telescope is in some measure wrong, since the Hubble Space Telescope has proved that many more stars are visible from space than are apparent from the ground.

Another explanation put forward to explain Olber's paradox is the Big Bang. More specifically, the extremely energetic radiation from the Big Bang has been red-shifted to microwave wavelengths (1100 times the length of its original wavelength) as a result of the cosmic expansion, and thus forms the cosmic microwave background radiation. This explains the relatively low light densities and energy levels present in most of our sky today despite the assumed bright nature of the Big Bang. Yet if examined in detail it is found that even this argument fails as an explanation of Olber's paradox.

Confusion about Olber's paradox Arises from discrepancies in Main stream theories on the nature of light:

Much of the modern confusion about Olber's paradox arises from the mainstream theory of light which states that a photon once emitted by an electron, will travel in a straight line forever until it is absorbed. Even a very simple geometrical calculation demonstrates that this is not a valid argument. It is common knowledge that particles radiating from a point source will diverge to an extraordinary degree. The photons composing a beam of light would diverge to such an extent that if they originate at some distant point (star) they would miss whole galaxies together as they diverge more and more and regress further and further away from the original cluster of photons that had originated at the source. In practise this is not what is observed. When light from a star is observed it is found to follow the inverse square law of distribution. That is the identical values for intensity can be found wherever the light is present and that it is present over an area varying directly with the square of the distance from the source. Therefore there is something very flawed in present day theories that light propagates as photons, which according to this theory are discrete and particle like. What is in fact observed is that photons propagate not as particles but as waves. Yet there is no doubt that photons also display particle like behaviour.

A photon is the smallest discrete amount or quantum of electromagnetic radiation. It is the basic unit of all light. Photons are always in motion and, in a vacuum, travel at a constant speed to all observers of 2.998×10^8 m/s. This is commonly referred to as the speed of light, denoted by the letter c. The nature of light — whether you regard it as a particle or a wave — was one of the greatest scientific debates. For centuries philosophers and scientists have argued about the matter that was (supposedly) barely resolved a century ago.

Theories of light prevailing during the eighteenth Century

The French scientist Descartes believed that an invisible substance, which he called the plenum, permeated the universe. Just as a wave travels through water, Descartes believed that light was a disturbance of the plenum that travelled through the plenum in the same way that a wave travels through water. Huygens believed that the universe was permeated by an ether, he surmised that the ether vibrated in the same direction as light, and formed a wave itself as it carried the light waves. In a later volume entitled, Huygens' Principle, he ingeniously described how each point on a wave could produce its own wavelets, which then add together to form a wave-front.

The Double Slit Experiment

Eventually, scientists had split into two entrenched camps. One side believed that light was a wave while the other view was of light as particles or corpuscles. The great champion of the so-called 'corpuscularists' was none other than Isaac Newton, widely believed to be the greatest scientist ever. Newton wasn't fond at all of the wave theory since that would mean light would be able to stray too far into the shadow. For much of the 18th century, corpuscular theory dominated the debate around the nature of light. But then, in May 1801, Thomas Young introduced the world to his now famous two-slit experiment where he demonstrated the interference of light waves. For the next century or so it was the wave theory of light that was dominant. This was the situation that prevailed until Max Planck's discovered that light (energy) was discrete in nature and that it consisted of tiny, discrete, indivisible, massless, electrically neutral, particle like packets of energy that he termed quanta. This could be considered to be the birth of quantum mechanics.

The quantum mechanics version of the Double Slit Experiment

Since the double slit experiment occupies such an important place in foundational quantum mechanics, it should be examined in more detail. Quantum Mechanics claims that a photon or any particle can be can be in two places simultaneously because of wave-particle duality and super position. The Double Slit Experiment as conducted with single photons or with electron or neutron or protons is offered as physical proof in support of this statement. What is the double slit experiment? Originally it was an experiment designed by Thomas Young to demonstrate that light underwent interference and this was definitely a wave attribute and not a particle attribute as claimed by Newton. The apparatus consisted of a screen with two narrow slits let into it which could be covered either singly or together, behind the screen was another screen on which the light passing through the screens could be seen, in front of both these screens was a light source. Thomas Young found that when only a single slit was left open and the light was shone through it, what was seen was a single large blob of light, that underwent diffraction, this is what was expected. No surprises here:

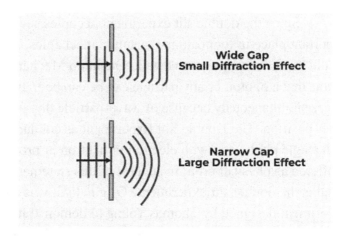

Diffraction

The light had undergone diffraction. Then both the slits were opened and it was found that the diffraction pattern disappeared, what was expected was either a larger blob of light or two diffraction patterns. This is only natural as with both slits open more light is available, so the expected pattern that light created when both slits were open was a double diffraction pattern as seen below:

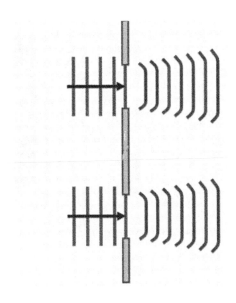

What was expected were two diffraction patterns or blobs of light

This is the pattern one would expect to find if light were a particle. Instead of this pattern what was found was this interference pattern of alternating light and shaded areas.:

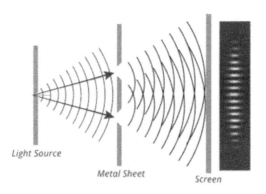

Interference pattern as seen in Double Slit experiment

The alternating light and dark pattern shown on the screen is the result of two waves interfering, a stream of particles could not under any circumstances form such a pattern. At the time it was thought that the Thomas Young Experiment was the ultimate proof that light was a wave.

With the advent of quantum mechanics and after Max Planck had discovered that light is not continuous but exists as minute discrete packets of energy called quanta or photons, the Double Slit Experiment was quoted as the Ultimate proof of the reality of quantum mechanics. Why? Here is where things get really strange. As technology became ever more sophisticated it was found that it was possible to send individual particles through the double slit experiment. We will call this phase the **Double Slit Experiment Phase A:** The result was strange not to say bizarre. When only one slit was open what was found was the expected diffraction pattern, remember that these were single particles and not waves. It was when single particles were sent in Indian file towards the screen with both slits open that weird effects began to be seen. If the experiment was allowed to run for a sufficient amount of time what was seen on the screen was not a brighter diffraction pattern or even two separate and distinct diffraction patterns but an interference pattern. Think about how mind boggling this result is! These are solid particles possessing a measurable mass, such as protons, electrons and neutrons that were behaving not as particles should do but as a wave. This behaviour, when one slit was open a diffraction pattern, when both slits were open an interference pattern, is considered to be the underlying proof of quantum weirdness and quantum logic. The question is raised as to how an individual particle could know that both slits were open? How did the particle know not to go to certain areas when only one slit was open? How did this happen? One of the conjectures of quantum mechanics was that perhaps an individual particle was able to pass through both slits at once (i.e., be in two places at the same time) in a process called superposition! Or perhaps the particle knew in some weird precognitive way that both slits were open. Here are some quotes from pre-eminent quantum mechanics physicists:

"The central mystery of quantum theory," wrote Henry Stapp, is 'How does information get around so quick how does the particle know that there are two slits? How does

the information about what is happening everywhere else get collected to determine what is likely to happen here?"

And

"Consciousness may be associated with all quantum mechanical processes . . . since everything that occurs is ultimately the result of one or more quantum mechanical event, the universe is "inhabited" by an almost unlimited number of rather discrete conscious, usually non-thinking entities that are responsible for the detailed working of the universe."

At this point you are thinking, things could not get more exotic and weird than this, well, things do indeed get more bizarre. We will call this phase the **Double Slit Experiment Phase B:** A natural consequence of seeing particles behaving like waves was that physicists tried to think up means by which particles could be detected at each slit, then some idea could be gained of how the interference pattern was created. So the experiment was repeated with detectors set up at each slit:

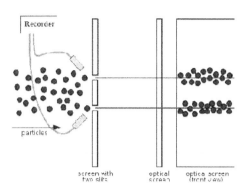

Double slit experiment with detectors set up at each slit

This is when something really strange begins to take place. As soon as the photo-detectors were put in place the interference pattern disappeared and a diffractions pattern was built up over time. Why did this happen? Could it be that the photons in some way knew that they were being observed and altered their behaviour accordingly? Could it be that at some mysterious sub-atomic level humans were not allowed to proceed beyond a certain point of awareness? It certainly was very strange that when no detectors were present and both slits were open an interference pattern was built up but with the detectors present the interference pattern disappeared and a diffraction pattern was seen, just as if the proton, electrons and neutrons were particles and not waves! What do the detectors show? If the experiment is properly controlled and particles are sent to the screen with the double slits individually, a subsequent particle being released only when the preceding particle has gone through the slit. It turns out that without exception only one detector, either left or right is triggered, both detectors are never triggered simultaneously. But then again when the detectors are present no interference pattern is seen! Weird. The Double Slit Experiment does indeed seem to be the ultimate conformation of the weird postulates of quantum mechanics; take wave particle duality and quantum superposition for instance.

However, for detractors of Quantum Mechanics this experiment is the ultimate proof of something else altogether. Here is the aether explanation of how the Double Slit Experiment works. According to Gestalt Aether Theory, the whole of space and indeed the Universe itself is permeated by a medium that allows light (electromagnetic radiation) to propagate. According to GAT this medium itself consists of light or virtual photons of such low energy (10^{-51}) that they are completely permeable to matter. Looked at from this point of view it is simple to deduce what happens. When the Aether (medium) encounters the double slits in an open state it passes through both slits and creates an interference pattern! If you are to understand the gist off this article it is important point to remember that the aether itself undergoes interference when it passes through the two slits. Even though, the aether has very low to zero interaction with matter, it is aware if it is passing through solids or a vacuum, hence the formation of an interference pattern when it passes through the two open slits.

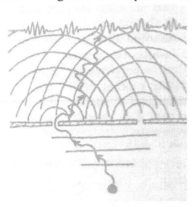

Photon or other particle creating an interference pattern as it passes through both open slits in the Double slit experiment.

The particles that are subsequently released merely follow the pattern that the Aether has already made and create an interference pattern on the screen. What could be more natural? Further this explains the results in the **Double slit experiment Phase B**: How does it do this? Take into account the fact that the only way to detect sub-atomic particles such as neutron or alpha particles, is through a photo-detector. In order to work the photo-detector has to create a field of its own, when this field is disturbed it means a particle has been detected. But when this field is disturbed it also means that the interference pattern created by the Aether is also disturbed since the

Aether might undergo polarization when the photo-detectors activate their fields. In any case the interference pattern disappears and the particles go straight through and create a diffraction pattern. Since particles are only detected going through one slit, either one or the other, this explanation of the disappearance the interference pattern when detectors are in place makes sense.

In this version of the experiment it is assumed that an undetectable aether exists, undetectable to the human sense but very apparent to micro particles. Since the aether exists everywhere, it follows logically that when both slits are open that the aether will travel through both slits and manifest as an invisible interference pattern. If single particles are now released they will follow as illustrated in the diagram above the path of the interference pattern created by the aether. When only one slit is open the aether will make a diffraction pattern and this is what the` micro particles will show. If both slits are open the aether will make an interference pattern which will be followed by the micro particles. Finally if the interference pattern created by the aether is disturbed as happens when detectors are in place, the interference pattern disappears and a diffraction pattern is seen. Thus the Double Slit Experiment can be taken as the ultimate proof of the existence of an aether, rather than as a proof of the esoteric quantum mechanics belief that particles are sentient, or that they can be in two places at once.

As can be seen from the above discussion of the double slit experiment, the more closely that quantum mechanics is examined, the more irrational and unreasonable it is found to be, while on the other hand the Gestalt Aether theory continues to strengthen its case for the existence of a Universe the background fabric of which is the aether.

Problems of the modern day theory of light

Thus the modern theory of light has many problems. It assumes that at the scale of the very small or sub-atomic, that particles can possess the properties of a solid, as for instance particles that possess mass and at the same time under other circumstances be massless and wavelike. The major problem with all these theories is that they assume that light travels for ever, even though it is assumed that the intensity of the light decreases inversely with the square of the distance from the source. The logical ending to such a sequence of events is that eventually the disturbance, in this case a wave of light, should become one with the ambience. Olber's paradox argues that if light were to travel fore ever, it would eventually accumulate, until the whole Universe would be as constantly as bright as the sun. This obviously does not happen the sky remains dark. How does modern science explain this circumstance?

Is Olber's Paradox the ultimate proof that modern theories on light are wrong?

A far more important aspect of Olber's paradox is that it completely undermines modern main stream theories of physics such as quantum mechanics and Standard Theory, and gives the lie to some of the most cherished notions in modern day mainstream physics. One of the idee fixes of quantum mechanics and standard theory is that:

"A photon once emitted by an electron will travel forever until it meets a solid object and is absorbed."

"Light just keeps going and going until it bumps into something. Then it can either be reflected or absorbed. Astronomers have detected some light that has been traveling for more than 13.3 billion years, close to the age of the universe." (Lee H, University of Illinois)

As you can see, light travels for a very long time if nothing gets in its way. We know from experiments that energy is always conserved, so unless there's something to which light can transfer its energy — and there's not much in deep space to interfere with— it will keep going forever. Light is made up of particles called photons that travel like waves. Unless they interact with other particles (objects), there is nothing to stop them.

Light can be invisible in three basic ways:

1) Light can be absorbed by obstructions in its way.
2) Light can be so far away that it has not reached.
3) Light can be red shifted out of the visible range by its source moving away at speed

According to Gestalt Aether Theory (GAT) this is not so. Instead, GAT holds that light behaves like any other physical process found in nature. The distance that light can travel, according to this theory, is based solely on the power involved in its production. This is the reason that a relatively weak source of light travels for only a limited distance. According to GAT –photons are produced through emission by electrons. These photons travel in straight lines or rays made up of hundreds of trillions of identical photons. As the lead photon comes into contact with the virtual photons of the aether that surround it, it passes on the whole of its energy, which is immediately replenished from the line of real photons behind it. The virtual photon that has acquired energy is thereby **promoted** to the status of a real photon and in turn shares its energy, promoting more virtual photons to the status of real photons and so on. Photons that have shared their energy, have that energy immediately replenished from the line of real photons. This is how light spreads out according to the inverse square law, while enabling individual photons to maintain their original energy Identity) as at the time of emission. When enough photons are not being generated the light can travel only a limited distance before the process of a continuous distribution of the original photon energy can no longer be sustained, resulting in a collapse of the wave which returns to the virtual photon state.

Consider that this makes a lot of sense. For instance it explains how the Hubble Space Telescope (HST) is able to see stars that had previously been invisible. For telescopes on earth, much of the light from distant stars in absorbed by the atmosphere making them invisible. Once the HST was launched into space the obstruction disappeared and distant stars became visible.

Yet the problem is much more complicated than just stating that light is absorbed by obstructions in its way. What are the properties of light in these circumstances? Take a bit of coal. When the coal is cold it doesn't emit any heat. When it is heated lightly, it glows red but this glow does not carry far. When the coal is heated further it turns into a luminous yellow that is visible from greater distances than was the red glow, lastly if the piece of lighted coal is placed in a stream of oxygen, the piece of coal burns with a bright white light that is visible from quite far away. Yet, however we burn the coal, we do not expect light from it to reach a distant star.

This observation that it is probable that light from the burning coal does not reach a distant star in turn raises other observations. How long the light burns for seems to make a difference. Lighting a match produces a momentary light, but does that light reach a distant star? So the criteria for how far light travels, depends on how much power is used to produce the light. Again, how far the light travels is also dependent to a certain extent for how long the light is on.

These observations raise another conundrum. What if you have a weak source of light that is on for a long time? Would that weak source of light be visible from a distant star? Could the light travel so far? Suppose you had an ordinary small bulb of 1W power, it was powered by a long lasting battery, how far do you think that light could travel? Would it travel far enough to reach a distant star? Almost definitely not, instead it would illuminate a fixed area further than which it cannot cross, using the power from the source to sustain illumination over a limited area. These are just ideas based on logic, but it should be remembered that light has to cross huge distances in its journey across the Universe, which means that it has to have huge power sources that last for a long time. A far more important aspect of Olber's paradox is that it completely undermines modern main stream theories of physics such as quantum mechanics and Standard Theory, and gives the lie to some of the most cherished notions in modern day mainstream physics. One of the idee fixe explanations for Olber's Paradox that is very much in favour in the present day, is the theory that the Universe is expanding at a rate that is much faster than the speed of light. How true can this be? Strictly speaking this should not be true, since according to Einstein, nothing can move faster than the speed of light, this postulate has to be true for the Universe as we know it to exist. Without the limiting speed of light, events could take place in any order and causality would not exist, resulting in chaos and a meaningless existence for the Universe. The explanation that is given for the faster than light expansion is that it is space itself that is expanding and therefore it is a cosmological expansion that is not affected by and does not affect the constancy of the speed of light.

Edwin Hubble and the Discovery of the Hubble Constant

The Hubble constant, a fundamental parameter in cosmology, measures the rate at which the universe is expanding. Its discovery revolutionized our understanding of the cosmos and underscored the dynamic nature of the universe. The constant is named after Edwin Hubble, the astronomer whose meticulous observations and dedication led to this ground breaking discovery. Edwin Powell Hubble was born on November 20, 1889, in Marshfield, Missouri, USA. Hubble's father wanted Edwin Hubble to be a lawyer which was contrary to Hubble's desire to study astronomy. Despite his father's wishes, Hubble pursued his passion for astronomy and made significant contributions to the field, including the discovery of the expanding universe. He showed an early interest in science and excelled academically, earning a scholarship to the University of Chicago, where he studied mathematics and astronomy. Hubble then attended the University of Oxford as a Rhodes Scholar, initially studying law before returning to his true passion, astronomy. He completed his Ph.D. at the University of Chicago in 1917.

After serving in World War I, Hubble joined the staff at the Mount Wilson Observatory in California. There, he had access to the 100-inch Hooker telescope, the most powerful telescope in the world at that time. This instrument would become crucial in his future discoveries.

Discovery of the Expanding Universe

In the 1920s, the prevailing view was that the Milky Way galaxy constituted the entire universe. However, Hubble's observations challenged this notion. By studying the light from distant "nebulae," Hubble discovered that these objects were not within the Milky Way but were separate galaxies far beyond our own. Hubble meticulously measured the distances to these galaxies using Cepheid variable stars as standard candles. These stars have a well-defined relationship between their luminosity and pulsation period, allowing for accurate distance calculations. Cepheid variables were discovered by Henrietta Swan Leavitt. In 1908, she found a relationship between the luminosity and the period of Cepheid variable stars, known as the period-luminosity relation, which later became a crucial tool for measuring astronomical distances. In 1924, Hubble published his results, demonstrating that the Andromeda Nebula was actually a separate galaxy, thus proving that the universe was much larger than previously thought. Further observations revealed another profound insight. Hubble noticed that the light from distant galaxies was red-shifted, indicating that they were moving away from us. This redshift increased with the galaxy's distance, suggesting that the universe was expanding. In 1929, Hubble formulated what is now known as Hubble's Law: the velocity of a galaxy is directly proportional to its distance from us. This relationship is expressed by the Hubble constant (H_0), which quantifies the rate of expansion of the universe. In order to measure the rate of recession of stars Hubble used the standard Doppler shift equations, since the relative Doppler shift was unknown at the time.

Dedication and Reliability of Hubble's Results

Edwin Hubble's dedication to his work was evident in his meticulous approach to observation and data analysis. He spent countless nights at the Mount Wilson Observatory, often enduring harsh conditions to gather the necessary data. Hubble's methods were rigorous, and his results were cross-verified with multiple observations and techniques, underscoring the reliability of his findings. Hubble's determination to unravel the mysteries of the cosmos was driven by a profound curiosity and a passion for discovery. His contributions laid the foundation for modern cosmology and opened new avenues of research, including the study of the Big Bang theory and the ultimate fate of the universe. Hubble's work earned him numerous accolades both during his lifetime and posthumously. He received the Gold Medal of the Royal Astronomical Society in 1935 and the Medal of Merit from the United States government in 1946 for his contributions to astronomy. Despite these honors, Hubble never received the Nobel Prize, as astronomy was not considered a part of physics by the Nobel Committee at that time. Edwin Hubble passed away on September 28, 1953, in San Marino, California. His legacy endures through the Hubble Space Telescope, launched in 1990 and named in his honour. This powerful instrument continues to explore the universe, providing breath-taking images and invaluable data that further our understanding of the cosmos.

The discovery of the Hubble constant by Edwin Hubble was a milestone in the field of astronomy. Hubble's work revealed the true scale and dynamic nature of the universe, fundamentally changing our perception of the cosmos. His dedication, meticulous methodology, and pioneering spirit established a new era in astronomical research, inspiring generations of scientists to explore the vast universe.

Georges Le Maitre and the Big Bang Theory

Georges Lemaître's pioneering work laid the theoretical groundwork for what would later be known as the Big Bang theory. In the 1920s, Lemaître, a Belgian priest and physicist, applied Einstein's general theory of relativity to the entire universe. He was deeply interested in the implications of Einstein's equations for cosmology, which suggested that the universe could be expanding. In 1927, Lemaître proposed that if the universe were expanding, it must have originated from a very small, dense state, which he called the "primeval atom" or "cosmic egg." This idea was revolutionary, as it suggested that the universe had a beginning—a moment of creation. Lemaître's conclusion stemmed from his examination of Einstein's equations, specifically the solutions that indicated an expanding or contracting universe. At the time, Einstein had introduced the cosmological constant (Λ) to allow for a static universe, which was the widely accepted view. However, Lemaître demonstrated that without the cosmological constant, the equations naturally led to an expanding universe. He suggested that if we reverse the expansion, we reach a point where the universe was a single, incredibly dense and hot point—a singularity. This marked the theoretical inception of what we now call the Big Bang.

Lemaître's theory gained observational support from Edwin Hubble's work. In 1929, Hubble, an American astronomer, provided empirical evidence for the expanding universe. By observing the redshift of light from distant galaxies, Hubble discovered that galaxies were moving away from us, and the further away they were, the faster they were receding. This relationship, now known as Hubble's Law, quantified the rate of expansion with what we call the Hubble constant (H_0). Hubble's findings substantiated Lemaître's theoretical framework by providing concrete evidence of an expanding universe, lending credence to the idea that the universe had a beginning point—the Big Bang. Hubble's discovery transformed our understanding of the cosmos. By measuring the velocities of galaxies and their distances, Hubble established that the universe was not static but dynamic and ever-expanding. This empirical data supported Lemaître's earlier theoretical work and helped shift the scientific consensus towards the Big Bang model of the universe's origin.

Building on these foundational ideas, we can further explore the implications of the Hubble constant as a kind of cosmic time machine. The Hubble constant measures the current rate of expansion of the universe. By looking back in time, we can infer that this rate was higher in the past. As we observe galaxies receding from us, we are effectively observing the universe at earlier stages of its expansion. This allows us to trace the history of the universe back to its origin. Extrapolating the Hubble constant back in time, we reach a point where the universe's size approaches zero—the moment of the Big Bang and the speed of recession nears the speed of light.

The further back we go, the faster the universe would have been expanding. This aligns with the inflationary model, which describes a brief period of exponential expansion immediately following the Big Bang. By treating the Hubble constant as a time machine, we can envision a scenario where, at the moment of the Big Bang, everything in the universe was moving at or near the speed of light. This initial rapid expansion set the stage for the more gradual expansion observed today.

In contrast to Lemaître's approach, which used Einstein's cosmological constant as a starting point, the present extrapolation uses the Hubble constant to trace the universe's history. Both methods ultimately point to a universe that began from a singularity, but they arrive at this conclusion through different means. Lemaître's theoretical insights provided the initial framework, while Hubble's empirical data substantiated the expanding universe. The extrapolation made in this book using the Hubble constant highlights the dynamic nature of cosmic expansion and allows us to visualize the universe's evolution from the Big Bang to its current state.

In summary, Georges Lemaître's theoretical work and Edwin Hubble's observational discoveries together revolutionized our understanding of the universe's origin and expansion. By treating the Hubble constant as a time machine, we can trace the universe's expansion back to the Big Bang, offering a profound glimpse into the dynamic history of the cosmos. This synthesis of theory and observation underscores the interconnectedness of time, space, and the fundamental nature of the universe, providing a comprehensive framework for understanding its evolution.

Inflation and Superluminal Expansion

The concept of inflation in the early universe provides a framework for understanding superluminal expansion. Inflation theory, proposed by Alan Guth in the 1980s, posits that a fraction of a second after the Big Bang, the universe underwent a period of exponential expansion. During this brief epoch, the universe expanded much faster than the speed of light. This rapid expansion helps to explain several observed properties of the universe, such as its large-scale homogeneity and isotropy. During inflation, the universe expanded by a factor of at least 10^{26} in a tiny fraction of a second, often calculated as being between 10^{-32} s and 10^{-36} s. This exponential growth was driven by a high-energy scalar field, often referred to as the inflation field. Post-inflation, the universe continued to expand, but at a much slower rate, eventually leading to the Hubble expansion we observe today.

In an attempt to try to build upon Alan Guth's hypothesis of an inflationary Big Bang, modern cosmologist try to suggest that this inflationary period the Big Bang can still be detected today and that the modern Universe is expanding at a super luminal speeds. This theory is far from tenable, to imagine that the effects of a time period of only 10^{-36} s could still be detectable today and that the modern Universe is expanding at these rates is untenable.

The concept of inflation in cosmology, first proposed by Alan Guth, suggests that the universe underwent a rapid exponential expansion in the first fractions of a second after the Big Bang. This period, lasting about 10^{-36} seconds, is thought to have driven the universe to expand at speeds far exceeding the speed of light, smoothing out any irregularities and leading to the homogeneous and isotropic universe we observe today. The theory was devised to address several puzzles in the standard Big Bang model, such as the horizon and flatness problems, and has become a central pillar of modern cosmology. However, the idea that the effects of this incredibly brief inflationary period could still be detectable billions of years later and that the universe continues to expand at superluminal speeds today raises significant challenges. The notion that a process lasting a mere 10^{-36} seconds could have such a profound and lasting impact on the entire cosmos is difficult to reconcile with our understanding of cause and effect on cosmic scales. While the inflationary model provides elegant solutions to certain cosmological problems, it also introduces complexities that are far from intuitive.

For instance, the idea of the universe expanding faster than the speed of light during inflation and continuing to do so in some interpretations of modern cosmology challenges our basic understanding of space and time. Though general relativity allows for the expansion of space itself at superluminal speeds, this concept is difficult to grasp and seems to stretch the limits of what we consider physically possible. The effects of such an expansion are not easily observable, and the indirect evidence supporting inflation, such as the uniformity of the cosmic microwave background radiation, does not provide direct proof of these extreme conditions.

Moreover, the idea that the universe is still expanding at such rates today, often referred to as "dark energy" driven expansion, relies heavily on interpretations of distant supernova observations and the redshift of galaxies. While these observations suggest an accelerating universe, the exact nature of this expansion and whether it indeed traces back to the inflationary epoch is still a topic of intense debate. The possibility that inflation and modern superluminal expansion are fundamentally connected remains speculative. While the inflationary Big Bang theory has been influential in shaping our understanding of the early universe, the suggestion that its effects remain detectable today, and that the universe continues to expand at superluminal speeds, is far from unchallenged. These ideas push the boundaries of our comprehension of the universe and require assumptions that, though mathematically consistent, are not entirely tenable when considered from a purely physical perspective.

Chapter 7
Gravity

Gravity is one of the fundamental forces of nature, shaping the universe in profound and essential ways. It is the force that gives us our sense of up and down, governs the motion of celestial bodies, and keeps us firmly anchored to the Earth's surface. Despite being the weakest of the four fundamental forces—electromagnetic, weak nuclear, strong nuclear, and gravitational—its effects are pervasive and universally experienced.

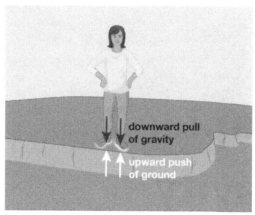

Gravity

From the grand scale of galaxies to the everyday experience of objects falling to the ground, gravity is an omnipresent force that dictates the structure and behaviour of the cosmos. The story of gravity begins with the observations of ancient civilizations, who noted the consistent behaviour of objects falling towards the Earth and the predictable movements of celestial bodies.

However, it was Sir Isaac Newton in the 17th century who provided a ground breaking understanding of gravity. In his seminal work, "Mathematical Principles of Natural Philosophy," Newton formulated the law of universal gravitation. He posited that every mass in the universe attracts every other mass with a force that is proportional to the product of their masses and inversely proportional to the square of the distance between their centres. This revolutionary insight explained not only why apples fall from trees but also why planets orbit the Sun, laying the foundation for classical mechanics. The one failing of Newton's Theory of Gravity was that although he was able to explain in great detail how gravity worked he was unable to explain the causative factor behind gravity, or why it worked in the way it did. He had to accept that his theory of gravity implied that gravity was a force that acted at a distance without any intervening medium. This is not to say that he was unaware of this failing, here is what he had to say:

In 1692, in his third letter to Bentley, he wrote: *"That one body may act upon another at a distance through a vacuum without the mediation of anything else, by and through which their action and force may be conveyed from one to another, is to me so great an absurdity that, I believe, no man who has in philosophic matters a competent faculty of thinking could ever fall into it."*

Newton also explained the fond hope that a solution would be found, he also left some pointers for future generations:

"Hitherto we have explained the phenomena of the heavens and of our sea by the power of gravity, but have not yet assigned the cause of this power. This is certain, that it must proceed from a cause that penetrates to the very centres of the sun and the planets, without suffering the least diminution of its force; that operates not according to the quantity of the surfaces of the particles upon which it acts (as mechanical causes used to do) , but

according to the quantity of solid matter which they contain, and propagates its virtue on all sides to great distances , decreasing always in the duplicate proportion of the distances ... ...But hitherto I have not been able to discover the cause of those properties of gravity from phenomena, and I frame no hypotheses (Hypotheses non fingo); for whatever is not deduced from the phenomena is to be called a hypotheses, whether metaphysical or physical, whether of occult qualities or mechanical, have no place in experimental philosophy. " Isaac Newton, Mathematica Principia.

Newton's law of gravity was immensely successful in describing the motion of planets and predicting the behavior of objects on Earth. However, it wasn't until the early 20th century that what was thought to be a more profound understanding of gravity emerged. Albert Einstein, with his theory of general relativity, proposed a radical new perspective. Instead of viewing gravity as a force between masses, Einstein described it as a curvature of spacetime caused by the presence of mass and energy. According to general relativity, massive objects like the Earth and the Sun warp the fabric of spacetime, creating the phenomenon we perceive as gravity. This curvature guides the motion of objects, causing them to follow curved paths. General relativity has been confirmed by numerous experiments and observations, such as the bending of light around massive objects and the precise calculations of planetary orbits.

Despite its foundational role in the cosmos, gravity remains an enigmatic force. It is extraordinarily weak compared to the other fundamental forces. For instance, the electromagnetic force between two electrons is about 10^{-40} times stronger than the gravitational force between them. This weakness means that gravity's influence becomes significant only when dealing with large masses like planets, stars, and galaxies. In our daily lives, we perceive gravity most directly through its effect on objects at or near the Earth's surface. It dictates the sensation of weight, the effort required to climb stairs, and the trajectory of a thrown ball. Gravity ensures that rivers flow downhill, that rain falls from the sky, and that we remain firmly planted on the ground. Gravity also plays a crucial role in the larger structure of the universe. It is the force that clumps matter together, forming stars, planets, and galaxies. In the vast emptiness of space, gravity is the architect that assembles matter into the complex structures we observe. It drives the formation of solar systems, guiding planets into stable orbits around their stars. On an even grander scale, gravity influences the dynamics of galaxy clusters and the large-scale structure of the cosmos. The mysterious dark matter, which makes up a significant portion of the universe's mass, reveals itself through its gravitational effects, despite being invisible to telescopes.

The study of gravity continues to be a vibrant field of research. Physicists strive to reconcile general relativity with quantum mechanics, seeking a unified theory that describes all fundamental forces. Gravity is a fundamental force that governs the behavior of the universe on both grand and everyday scales. From the falling of an apple to the motion of galaxies, gravity is the force that shapes the cosmos. It tells us what is up and down, dictates the orbits of planets, and keeps us grounded on Earth. While it is the weakest of the fundamental forces, its influence is universal and profound. As our understanding of gravity deepens through on- going research and discovery, we continue to unravel the mysteries of this essential force, gaining ever greater insights into the workings of the universe. Newton and Einstein both made ground breaking contributions to our understanding of gravity, but they conceptualized it in fundamentally different ways. Newton postulated a gravitational force, while Einstein introduced the idea of a gravitational potential. To address whether Einstein simply used gravity to explain gravity, we need to delve into the nuances of their theories and what they each sought to explain.

Newton's Gravitational Force

Newton's law of universal gravitation describes gravity as a force acting at a distance between two masses. This force is directly proportional to the product of the masses and inversely proportional to the square of the distance between them. Mathematically, this is expressed as:

$$F = \frac{G(m_1 m_2)}{(r^2)}$$

where F is the gravitational force, G is the gravitational constant, m_1 and m_2 are the masses of the objects, and r is the distance between the centres of the masses.

Newton's theory was immensely successful in explaining a wide range of phenomena, from the falling of an apple to the orbits of planets. However, it didn't address why gravity existed or how it acted instantaneously across vast distances. Newton himself was cautious about these aspects, famously remarking, "I frame no hypotheses" regarding the mechanism behind gravity.

Einstein's Gravitational Potential

Einstein revolutionized our understanding of gravity with his theory of general relativity. Instead of treating gravity as a force acting at a distance, Einstein described it as a curvature of spacetime caused by mass and energy. In this framework, massive objects like the Earth warp the spacetime around them, and this curvature dictates the motion of objects. Objects follow the geodesics, or the straightest possible paths, in this curved spacetime.

Einstein's field equations, which form the core of general relativity, relate the curvature of spacetime (expressed by the Einstein tensor $G\mu\nu$ to the energy and momentum of matter and radiation (expressed by the stress-energy tensor $T\mu\nu$:

$$G_{\mu\nu} = 8\pi G/c^4\, T_{\mu\nu}$$

In this context, the "gravitational potential" refers to the metric tensor $g_{\mu\nu}$, which encapsulates the geometry of spacetime. The metric tensor describes how distances and time intervals are measured in the presence of gravity.

Did Einstein Use Gravity to Explain Gravity?

Because Einstein merely replaced Newton's gravitational force by a gravitational potential, the accusation is often raised that Einstein merely used gravitation as an explanation of itself. Although this might seem to be true in actual fact Einstein's theory replaced the concept of a gravitational force with a geometric interpretation. He explained that what we perceive as the force of gravity is actually the result of objects moving along curved paths in a warped spacetime. This is a significant shift from Newton's action-at-a-distance force to a more intricate and underlying geometric property of spacetime itself.

By describing gravity as the curvature of spacetime, Einstein's theory could explain phenomena that Newton's theory could not. For example, general relativity accounts for the precise perihelion precession of Mercury's orbit and predicts the bending of light around massive objects, phenomena that were inexplicable by Newtonian mechanics.

General relativity also provides a framework that is consistent with the principle of equivalence, which states that locally (in a small enough region of spacetime), the effects of gravity are indistinguishable from acceleration. This principle underlies Einstein's geometric interpretation and shows how gravity can be understood as a property of spacetime rather than a separate force. In summary, Einstein did not use gravity to explain gravity in a circular manner. Instead, he redefined the very nature of gravity, moving from a force between masses to a manifestation of spacetime curvature. This shift provided a deeper understanding of gravitational phenomena and resolved inconsistencies in Newtonian mechanics, thereby offering a more comprehensive and elegant explanation of the universe's behaviour.

General relativity is a metric theory of gravitation. At its core are Einstein's equations, which describe the relation between the geometry of a four-dimensional, pseudo-Riemannian manifold representing space-time, and the energy–momentum contained in that space-time. Phenomena that in classical mechanics are ascribed to the action of the force of gravity (such as free-fall, orbital motion, and spacecraft trajectories), correspond to inertial motion within a curved geometry of space-time in general relativity; there is no gravitational force deflecting objects from their natural, straight paths. Instead, gravity corresponds to changes in the properties of space and time, which in turn changes the straightest-possible paths that objects will naturally follow. The curvature is, in turn, caused by the energy–momentum of matter. Paraphrasing the relativist John Archibald Wheeler : *space-time tells matter how to move; matter tells space-time how to curve.*

The Gestalt Aether Theory of Gravity

Gestalt Aether Theory presents a novel perspective on gravity, diverging from traditional interpretations found in general relativity, quantum mechanics and classical physics. In this theory, gravity is not a fundamental force but rather an emergent property arising from the interactions of electrons and the virtual photon aether (Dark Matter). It should be stated that Dark Matter is an excellent candidate for such a virtual photon aether. It represents 85% of all matter in the Universe, at a conservative estimate, it allows the free passage of types of electromagnetic radiation without offering the slightest interference, it has very low to zero interaction with matter, as predicted by Gestalt Aether Theory, it exerts a gravitational force as explained by Gestalt Aether Theory, In short, given these properties, it is difficult to understand what else Dark Matter might be. This conceptual framework shifts our understanding of gravity from a geometric or force-based phenomenon to one based on the

242

dynamic behaviours of subatomic particles. At the heart of Gestalt Aether Theory is the behaviour of electrons orbiting atomic nuclei. This model of the atom rejects the quantum mechanics assumption of the electron existing as a wave function or electron cloud within the atom and instead treats the electron as a solid particle that is orbiting the electron in definite energy orbitals. In a hydrogen atom, for instance, the electron completes approximately 2.59×10^{15} orbits around the nucleus every second. During each orbit, the electron emits and absorbs virtual photons, processes essential for the electron to stabilize its energy around the nucleus. These emissions and absorptions (self-interactions) are not random but occur in a highly ordered sequence, maintaining the electron's energy balance. The virtual photon emitted by an electron may possess the energy of a real photon, this is what causes the virtual photons of the virtual photon aether (Dark Matter) to align in the direction of propagation of the emitted virtual photon, since the virtual photon emitted by the electron and the virtual photons of the virtual photon aether (Dark Matter) are both electric dipoles, differing only in energy content, but the interaction (emission and absorption) takes place over such a short time $(10^{-18}$ s) that no energy is conveyed along the aligned line of virtual photons, instead the line of aligned photons, represents the shortest distance between two objects. In order to evaluate the implications of incorporating the universal photon aether(Dark Matter) into our understanding of gravity and compare it with

Einstein's general relativity, it is essential to first summarize the core principles of both Newtonian and Einsteinian gravity, and then explore how the concept of a photon aether might provide an alternative explanation.

Newtonian Gravity and Action at a Distance

Newton's law of universal gravitation was revolutionary in describing how objects with mass attract each other. However, it assumed an instantaneous action at a distance, which means that changes in the gravitational field would be felt instantly, no matter the distance. This notion was problematic because it implied a mechanism that could act faster than the speed of light, violating the principles of causality and the finite speed of information transfer. It should be understood that Newton himself was well aware of this problem, but rather than speculate, he left the problem to be solved by others with his famous quote "Hypotheses non fingo" : I make no hypotheses. Gestalt Aether Theory explains this criteria of 'action at a distance', by proposing Dark Matter as the medium through which both electromagnetic radiation and light propagate.

Einstein's General Relativity and Spacetime Curvature

Einstein's general relativity resolved the action-at-a-distance problem by introducing the concept of spacetime curvature. Instead of gravity being a force that acts instantaneously across a distance, Einstein described it as the warping of spacetime by mass and energy. Objects move along the curved paths in this warped spacetime, which we perceive as gravitational attraction. This framework inherently includes the finite speed of gravitational interactions, which propagate at the speed of light, consistent with the theory of relativity. However, the theory of relativity itself would cease to exist if a medium for the propagation of light is proved to exist, therefore if an aether like medium is proven to exist it is not necessary to consider either General relativity or special relativity in this context.

The idea of a universal photon aether involves a medium (dark matter) filled with virtual photons, which could potentially permeate every part of the Universe, just as suggested by Newton, that mediate gravitational interactions. This aether could serve as the underlying fabric through which gravitational effects are transmitted, providing a mechanism that doesn't require the instantaneous action at a distance that troubled Newtonian mechanics. Virtual photons, according to Gestalt Aether Theory, permeate the entire universe, forming what is known as the virtual photon aether or dark matter. These virtual photons are the medium through which electromagnetic radiation propagates and are integral to the mechanism of gravity in this theory. Each time an electron emits a virtual photon, the surrounding virtual photons, in line with the direction of propagation of the emitted photon form into a line. This alignment takes place since they are electric dipoles and align in response to the emitted photon, which is also an electric dipole and identical in structure to the virtual photons of the aether (Dark Matter). Unlike in the case of when a real photon is emitted and its energy is conveyed along the line of virtual photons, when a virtual photon is emitted the time period is extremely brief, about 10^{-18} s. this means that there is no energy flowing along the lines of force, it only results in their alignment for a very brief period. This alignment is not isolated but propagates through the virtual photon aether. This alignment of the virtual photon aether into a line of force, represents the shortest distance between two bodies, if the line of force lands on another body, the aligned line of force represents the shortest distance between the two objects, it also results in a proliferation of the lines of force between the two objects in keeping with the number of electrons available in the two objects and the resultant number of aligned lines of force it can produce, therefore, the proportional force with which the two objects are gradually pulled together is dependent on the number of lines of force each object can generate. If one of the objects is denser than the other, it will result in the generation of a greater number of lines of force and the lighter object would be pulled towards the heavier (denser) object. These alignments are crucial because they represent the collective behaviour of the virtual photon aether, giving rise to a macroscopic effect we perceive as gravity. The question that was asked at the beginning of this book as to why objects can be shielded from electromagnetic radiation but not from gravity can be answered, by stating that while electromagnetic radiation involves the exchange of energy, gravity is due to virtual interactions and do not register in the macro world.

Gravity as an emergent property of matter in Gestalt Aetyher Theory

In Gestalt Aether Theory, gravity emerges from the cumulative alignments of virtual photons around matter. When two masses are present, the virtual photon aether between them aligns due to the alignment of the virtual photon aether by the continuous emission and absorption of virtual photons by the electrons within each mass. This alignment creates a tensioning effect in the aether, representing the shortest path between the two objects. The virtual photon alignments act as a connective network, pulling the objects together with a force that we recognize as gravitational attraction. The strength of this gravitational force is proportional to the density of matter and inversely proportional to the square of the distance between the objects, aligning with the observed inverse-square law of gravity. This alignment creates a force that, while significantly weaker than the electromagnetic force (by a factor of 10^{-40}), is nonetheless pervasive and influential on a cosmic scale.

In essence, gravity in Gestalt Aether Theory is not a field or curvature of spacetime but a manifestation of the interactions between electrons and the virtual photon aether. The continuous emission and absorption of virtual photons by electrons create a dynamic network of alignments in the aether, producing an attractive force between masses. This view integrates the micro-level behaviours of particles with the macro-level phenomenon of gravity, offering a cohesive explanation that emerges from the fundamental properties of matter. By focusing on the electron's behaviour and its interaction with the virtual photon aether, Gestalt Aether Theory provides a unique lens through which to understand gravity. It posits that the gravitational force we observe is a direct result of the constant self-regulation of electron energy through virtual photon interactions. This framework not only redefines our concept of gravity but also highlights the intricate and interconnected nature of the universe at both microscopic and macroscopic levels.

Comparison between electromagnetic and gravitational forces as explained by Gestalt Aether Theory.

Comparing the effect of the propagation of electromagnetic radiation with gravity, it should be noted that electromagnetic radiation involves lines or rays of identical connected photons numbering 10^{14} per second, all travelling in the same direction and all with identical energies. The lines of force involved in gravity by contrast are not aligned in this manner and also convey no energy, this accounts for the huge difference in energy between the gravitational force and the electromagnetic force.

Gestalt Aether Theory presents gravity as an emergent property of matter, arising from the relentless alignment of the virtual photon aether caused by the electron's emission and absorption of virtual photons. This theory offers a fresh perspective, emphasizing the profound impact of subatomic interactions on the large-scale structure and behaviour of the universe. Through this lens, gravity becomes a tangible and comprehensible consequence of the fundamental processes governing the particles that make up all matter.

Newton's Theory of Gravity:

Isaac Newton, in his ground breaking work "Philosophiæ Naturalis Principia Mathematica," established the law of universal gravitation, positing that every mass exerts an attractive force on every other mass. This force is proportional to the product of their masses and inversely proportional to the square of the distance between them. Newton's formulation succinctly captures gravity's observable effects without delving into its underlying mechanism. He acknowledged gravity's action at a distance but refrained from hypothesizing about the force's transmission through space, focusing instead on its mathematical precision and universality. Gestalt Aether Theory (GAT) aligns with Newton's conceptualization of gravity's effects while providing a unique explanation for its cause. In GAT, gravity emerges from the interactions of electrons within atoms and the surrounding virtual photon aether, or dark matter. This theory posits that electrons continuously emit and absorb virtual photons to stabilize their energy. In response to the emission of a real photon by the electron, the virtual photons of the aether align in response to the electrons' emissions, creating a line of aligned photons within the aether. This alignment results in the shortest distance between two objects and manifests as an attractive force between masses, which we perceive as gravity. Unlike Newton's theory which had to assume that gravity acted at a distance, Gestalt Aether Theory posits a material substance Dark Matter or the virtual photon aether that acts as a medium for the gravitational force.

Similarities between Newtonian Gravity and the Gestalt Aether Theory of Gravity:

Both Newton's explanation and GAT agree on the inverse-square nature of gravitational attraction. Newton's law, $F = G\ (m_1\ m_2/r^2)$, describes gravity as diminishing with the square of the distance between two masses. Similarly, GAT describes gravity as resulting from the alignment of virtual photons, which also follow an inverse-square relationship due to the spatial propagation of these alignments. This similarity underscores the consistent observation that gravity weakens with distance. A key aspect of GAT is the role of density in the strength of gravitational attraction. In Newtonian mechanics, the gravitational force depends on the mass of objects, implicitly involving their density. GAT makes this dependence explicit by linking it to the number of electrons within an atom. More electrons result in more frequent emissions and absorptions of virtual photons, creating more lines of aligned force within the virtual photon aether. This increased density of alignments directly correlates with a stronger gravitational force, aligning with Newton's observation that larger masses exert stronger gravitational attractions.

In both theories, gravity emerges from the centre of masses and acts continuously and relentlessly over great distances. Newton's law applies universally, affecting celestial bodies and everyday objects alike. GAT echoes this universality, with the virtual photon alignments occurring throughout the universe, facilitating gravity's pervasive influence. Another parallel is the recognition of gravity as the weakest of the fundamental forces. In Newton's view, gravity is significant but weaker than electromagnetic forces, a distinction that GAT emphasizes by quantifying gravity as 10^{40} times weaker than the electromagnetic force. This weakness arises because gravity, in GAT, is a virtual force generated by the virtual interactions of photons, in contrast to the real interactions governing electromagnetic forces.

Gravity's attractive nature is a crucial similarity between the two theories. Newton's law describes gravity as always pulling masses together. GAT explains this by noting that when virtual photon lines of force align between two objects, it causes a reciprocal alignment in the virtual photon aether. This mutual alignment leads to an attractive force, ensuring that gravity is consistently attractive rather than repulsive. It could be said that the Gestalt Aether Theory of gravity ratifies or explains Newtonian gravity by providing a detailed mechanism behind the gravitational force that Newton described mathematically. The concept of gravity acting over great distances is also inherent in both theories. Newton's law accurately predicts the gravitational influences of distant celestial bodies, while GAT posits that the continuous alignments of virtual photons create a network that extends across vast expanses, allowing gravity to operate over immense distances.

Newton's explanation of gravity and the Gestalt Aether Theory share several key similarities. Both recognize gravity's inverse-square nature, its emergence from the centre of masses, its continuous and relentless action, and its universal applicability. GAT enhances our understanding by providing a mechanism for gravity rooted in the behaviour of electrons and virtual photons, aligning with Newton's empirical observations. By emphasizing the role of density, the weakness of the gravitational force, and its always-attractive nature, GAT complements Newton's mathematical framework with a conceptual model that elucidates the underlying cause of gravity. This synthesis of descriptive and explanatory elements offers a richer perspective on one of the universe's most fundamental forces.

Effect of validating an aether like medium

The foundational principles of special and general relativity are deeply embedded in modern physics. Special relativity, proposed by Albert Einstein in 1905, revolutionized our understanding of space and time by asserting that the speed of light is constant in a vacuum and that the laws of physics are the same for all observers in inertial frames of reference. General relativity, introduced by Einstein in 1915, extended these ideas to include gravity, describing it as the curvature of spacetime caused by mass and energy. However, if a medium were proven to exist through which electromagnetic radiation propagates as Gestalt Aether Theory posits, it would challenge the core assumptions of both special and general relativity. This medium, often conceptualized as the aether in historical contexts or as the virtual photon aether (Dark Matter) in the Gestalt Aether Theory, implies a preferred frame of reference for the propagation of light and other electromagnetic waves. Such a medium would fundamentally alter our understanding of the nature of light, space, and time.

The Gestalt Aether Theory posits that this medium, composed of virtual photons, permeates the universe and facilitates the propagation of electromagnetic radiation. In this theory, gravity emerges from the interactions of electrons with the virtual photon aether. Every time an electron emits and absorbs a virtual photon, it aligns the virtual photon aether, creating an attractive force between masses. This continuous and relentless alignment explains gravity as an emergent property of matter, challenging the geometric interpretation of gravity in general relativity. If the virtual photon aether is confirmed, as seems increasing likely, the constancy of the speed of light would no longer be a fundamental postulate, as the speed of light would depend on the properties of this medium. This would contradict special relativity, which relies on the invariance of the speed of light in a vacuum. The presence of a medium would suggest that light's speed could vary based on the medium's characteristics, introducing a preferred frame of reference and violating the principle of relativity.

Moreover, the concept of spacetime curvature in general relativity would be reinterpreted. Instead of gravity being a manifestation of curved spacetime, it would be viewed as a force resulting from the interactions within the virtual photon aether. This shift would necessitate a rethinking of gravitational phenomena, from planetary orbits to black holes, under the framework of the aether medium rather than spacetime geometry.

Such a paradigm shift would have profound implications for physics. It would require a re-evaluation of experimental evidence supporting relativity and the development of new theoretical models to describe the interactions within the virtual photon aether. Experiments designed to detect the presence of this medium and measure its properties would become crucial in validating or refuting the Gestalt Aether Theory. The implications extend beyond gravity. Electromagnetic phenomena, governed by Maxwell's equations, would also need reinterpretation within the context of the aether medium. The propagation of electromagnetic waves, their interactions with matter, and the nature of electromagnetic fields would be viewed through the lens of the virtual photon aether. This could lead to new insights into the nature of light, matter, and their interactions.

In conclusion, the confirmation of a medium for electromagnetic radiation, such as the virtual photon aether proposed in the Gestalt Aether Theory, would challenge the foundations of special and general relativity. It would

255

necessitate a paradigm shift in our understanding of gravity, light, and the nature of the universe. This new framework would offer an alternative explanation for gravity as an emergent property of matter, rooted in the interactions within the virtual photon aether, fundamentally altering the landscape of modern physics.

The Formation of Stars

Stars are born from vast clouds of gas and dust, known as nebulae, scattered throughout galaxies. These nebulae, primarily composed of hydrogen, the simplest and most abundant element in the universe, provide the raw material for star formation. The process begins when a region within a nebula experiences a disturbance—such as a nearby supernova explosion—that causes it to collapse under its own gravity. As the gas cloud collapses, it fragments into smaller clumps, each of which can become a star. As a clump collapses, it begins to heat up due to the conversion of gravitational potential energy into thermal energy. This process continues until the core of the clump becomes hot and dense enough to initiate nuclear fusion—the process that powers stars. When this occurs, a proto-star is born. In the core of a young star, hydrogen atoms undergo nuclear fusion, a process where they combine to form helium, releasing enormous amounts of energy in the form of light and heat. This energy counteracts the force of gravity, maintaining the star's stability in a state known as hydrostatic equilibrium. The specific fusion reactions depend on the star's mass:

In low to medium mass stars like the Sun, the primary fusion process is the proton-proton chain reaction, where four hydrogen nuclei (protons) combine through a series of steps to form one helium nucleus, releasing energy in the process. This reaction is efficient at the temperatures and pressures found in stars like our Sun.

In more massive stars, the core temperature is higher, allowing for a different fusion process called the carbon-nitrogen-oxygen (CNO) cycle. In this cycle, hydrogen is fused into helium with carbon, nitrogen, and oxygen acting as catalysts. This cycle produces energy more efficiently than the proton-proton chain at the higher temperatures found in massive stars. As hydrogen in the core is converted to helium, the star remains stable for millions to billions of years, depending on its mass. The more massive the star, the shorter its lifespan, as it burns through its nuclear fuel more quickly.

The Star's Evolution

Eventually, the hydrogen fuel in the star's core is depleted, leading to significant changes in the star's structure and behavior. What happens next depends largely on the star's mass. In low to medium mass stars that are up to eight times the mass of the Sun, as the hydrogen in the core is exhausted, the core contracts and heats up, causing the outer layers of the star to expand and cool, turning the star into a red giant. In this phase, helium fusion begins in the core, converting helium into carbon and oxygen. The outer layers may be ejected, forming a planetary nebula, while the core remains as a white dwarf.

The core left behind after the red giant phase is a white dwarf, a dense, Earth-sized remnant composed mainly of carbon and oxygen. With no more nuclear reactions to support it, the white dwarf gradually cools and fades over billions of years. It is prevented from further collapse by electron degeneracy pressure, a quantum mechanical effect that arises from the Pauli exclusion principle.

In massive Stars that are eight to twenty-five times the Mass of the Sun, the story is more dramatic. After exhausting their hydrogen and helium, these stars continue fusing heavier elements in their cores, forming layers of elements like carbon, neon, oxygen, and silicon. Once the core is primarily composed of iron, fusion can no longer produce energy, as fusing iron consumes energy rather than releasing it. The core becomes unstable and collapses under its own gravity, leading to a catastrophic supernova explosion. This explosion disperses the outer layers of the star into space, leaving behind a dense core.

If the core of the star that remains after the supernova explosion takes place is between 1.4 and 3 times the mass of the Sun , it becomes a neutron star. In this incredibly dense object, protons and electrons combine to form neutrons, resulting in a star composed almost entirely of neutrons. Neutron stars are only about 20 kilometers in diameter but have masses greater than the Sun, making them extraordinarily dense. The gravitational force is very intense. Neutron stars often spin rapidly, emitting beams of radiation as pulsars.

In the really massive stars that are over twenty-five times the Mass of the Sun there is a chance of the forming of a black hole. If the core left after a supernova explosion is more than three times the mass of the Sun, not even neutron degeneracy pressure can halt its collapse. The core continues to collapse, forming a black hole—a point in space where gravity is so strong that not even light can escape. The boundary surrounding a black hole is known as the event horizon, beyond which nothing can return. Black holes continue to attract nearby matter and can grow in size over time. They often become the centers of extreme gravitational environments and are key players in phenomena such as quasars and gamma-ray bursts.

The life cycle of a star is a complex interplay of nuclear reactions and gravitational forces, with the final fate of the star determined largely by its initial mass. From the delicate balance of fusion and gravity in a star's core to the dramatic events of supernovae and black hole formation, stars are both the creators and destroyers of the elements that make up the universe. Each stage of a star's life contributes to the cosmic tapestry, from the birth of new stars in nebulae to the enrichment of the interstellar medium with heavy elements, leading to the formation of planets and, ultimately, life itself. The death of a star, whether it ends as a white dwarf, neutron star, or black hole, is not just an end but a continuation of the cosmic cycle, as the remnants of these stars go on to influence the next generation of stars and galaxies.

Neutrinos:

The history of the neutrino is a fascinating tale of theoretical prediction, experimental perseverance, and the gradual unveiling of one of the most elusive particles in the universe.

In the early 20th century, physicists were grappling with a perplexing problem. During the study of beta decay—a process where an unstable nucleus emits an electron (or positron)—it was observed that the emitted electrons did not carry away all the energy expected from the decay. According to the conservation of energy, this discrepancy suggested that either the principle was being violated, or there was something else happening that could not be accounted for. In 1930, the brilliant physicist Wolfgang Pauli proposed a daring solution. He suggested that a previously unknown particle was being emitted alongside the electron during beta decay. This particle, Pauli theorized, was electrically neutral, nearly massless, and extremely difficult to detect, as it interacted only very weakly with matter. Pauli was initially hesitant to publish this idea because the existence of such a particle seemed almost impossible to verify experimentally. He referred to this hypothetical particle as the "neutron," though this term was later appropriated for the particle discovered by James Chadwick in 1932. To avoid confusion, Enrico Fermi later coined the term "neutrino," meaning "little neutral one" in Italian. Fermi's theory of beta decay, formulated in the 1930s, incorporated the neutrino as a fundamental component of the process. According to Fermi, during beta decay, a neutron in the nucleus transforms into a proton, emitting both an electron and a neutrino in the process. Fermi's theory was a major step forward, but it still left the neutrino as a purely theoretical entity.

The quest to detect the neutrino directly became a significant challenge for experimental physics. Because neutrinos interact so weakly with matter, they can pass through vast amounts of material without leaving a trace. It was not until 1956, over two decades after Pauli's initial proposal, that the neutrino was finally detected. Physicists Clyde Cowan and Frederick Reines conducted an experiment near a nuclear reactor, which produces a copious number of neutrinos. By observing the reactions these neutrinos occasionally caused in a large tank of water, Cowan and Reines provided the first direct evidence of the neutrino's existence. Their work was later recognized with the Nobel Prize in Physics.

Following its discovery, the neutrino became an object of intense study. It was soon realized that there were multiple types, or "flavors," of neutrinos corresponding to the different types of charged leptons: the electron neutrino, the muon neutrino, and the tau neutrino. These different flavors were identified over the subsequent decades as particle accelerators and detectors became more sophisticated. One of the most intriguing aspects of neutrinos is their mass—or rather, the lack thereof. For many years, neutrinos were believed to be massless, as assumed by the Standard Model of particle physics. However, in the late 20th century, experiments studying neutrinos from the sun and cosmic sources suggested that neutrinos undergo a process known as "oscillation," where they change from one flavor to another as they travel. This phenomenon could only be explained if neutrinos had a small, but nonzero, mass. This discovery, which earned the 2015 Nobel Prize in Physics, was groundbreaking because it showed that the Standard Model was incomplete and hinted at new physics beyond what was previously known.

Today, neutrinos continue to be at the forefront of research in particle physics and cosmology. They are key to understanding processes ranging from the behavior of supernovae to the evolution of the early universe. Despite their elusive nature, neutrinos hold the potential to answer some of the deepest questions in physics, making their history not just a story of discovery, but of ongoing exploration into the fundamental nature of reality.

Neutrinos have long intrigued scientists due to their elusive nature and the significant role they play in our understanding of the universe. Traditionally, neutrinos are considered elementary particles, virtually massless and chargeless, emerging from high-energy processes such as nuclear reactions in stars, supernovae, and radioactive decay. They interact extremely weakly with matter, passing through it almost undetected, which has led to their reputation as 'ghost particles.' However, the Gestalt Aether View offers a radically different interpretation of neutrinos, suggesting that what we consider neutrinos may, in fact, be manifestations of a more fundamental process involving the virtual photon aether.

Gestalt Aether Theory and neutrinos

According to the Gestalt Aether theory, neutrinos are not particles in the traditional sense. Instead, they are the end result of the destruction of a nucleus, which triggers the emission of an energetic virtual particle. This event causes an alignment of the virtual photon aether, forming what we perceive as a line of force. The theory posits that it is this line of force, rather than an actual particle, that we mistakenly identify as a neutrino. This reinterpretation implies that neutrinos are not independent particles but rather the result of complex interactions within the virtual photon aether, which itself is a vast, omnipresent field of virtual particles that permeates the universe.

The destruction of a nucleus is a violent process, often associated with tremendous energy release, whether in a supernova explosion, the fusion reactions powering the Sun, or the fission processes in nuclear reactors. When a nucleus is destroyed, it does not merely disintegrate into smaller particles but also results in the release of virtual particles resulting in the alignment of the virtual photon aether. This creation of lines of force are regions where the virtual photons align in response to the energetic event. These lines of force, with their varying energies, are what we interpret as neutrinos. Neutrinos are known to come in different "flavors" or types—electron, muon, and tau neutrinos—each associated with different processes and energies. The Gestalt Aether View suggests that these different types correspond to different configurations or energies of the lines of force created during the destruction of nuclei. This would mean that neutrinos of different energies are not separate entities but are variations in the alignment of the virtual photon aether, each type reflecting the specific conditions of the nuclear interaction that produced it.

The Sun and the stars as sources of neutrinos

Because of the fact that the stars exist through a process of nuclear fusion, they are a perpetual source of neutrinos. The Sun is a prolific source of neutrinos, with approximately 10^{10} neutrinos passing through every square centimeter of our bodies each second. Despite their enormous numbers, these neutrinos interact so weakly with matter that they have no observable effect on us.

According to the Gestalt Aether View, this is because the lines of force representing neutrinos are not sufficiently concentrated to influence matter on a macroscopic scale. However, in extreme environments, such as those found in neutron stars, the situation changes dramatically.

Neutron stars

Neutron stars, the remnants of massive stars that have undergone supernova explosions, are incredibly dense, and their gravity is so intense that it warps spacetime itself. In these stars, the density of neutrinos—or more accurately, the density of the lines of force representing neutrinos—reaches levels where they can significantly influence the star's structure and dynamics. The Gestalt Aether View suggests that in neutron stars, the concentration of these lines of force becomes so high that they generate a form of gravity distinct from ordinary gravity. This extraordinary gravitational force could be responsible for the collapse of the star into a black hole, where even light cannot escape its pull.

In this framework, black holes could be seen as regions where the virtual photon aether has become so densely aligned, due to the intense production of neutrinos and their associated lines of force, that it creates a gravitational well from which nothing can escape. This gravity, born from the alignment of the virtual photon aether, would be an emergent property of the interactions taking place within the dense environment of the neutron star, leading to the formation of a black hole.

The Gestalt Aether View thus presents a fascinating reinterpretation of neutrinos, not as particles but as manifestations of deeper, more fundamental processes within the virtual photon aether. This perspective challenges the conventional understanding of neutrinos and opens new avenues for exploring the nature of matter, energy, and the fundamental forces that govern the universe. While this theory remains speculative and diverges significantly from established scientific models, it offers a thought-provoking alternative that invites us to reconsider the true nature of the particles and forces that shape our reality.

To delve deeper into the formation and behavior of neutrinos, it's essential to first understand their origins within the standard model of particle physics. Neutrinos are primarily produced in processes involving the weak nuclear force, one of the four fundamental forces of nature. This force governs the behavior of subatomic particles and is responsible for processes like beta decay, a type of radioactive decay where a neutron transforms into a proton, an electron, and an electron antineutrino.

Formation of Neutrinos

Neutrinos are formed in various high-energy processes, particularly within stars, supernovae, and certain types of radioactive decay. The most common example is within the Sun, where nuclear fusion reactions convert hydrogen into helium. During these reactions, protons combine to form deuterium (a hydrogen isotope), and in this process, a positron and an electron neutrino are emitted. This neutrino is a byproduct of the weak nuclear force, specifically during the interaction that converts a proton into a neutron. In the case of beta decay, which occurs in radioactive isotopes, a neutron in the nucleus of an atom decays into a proton, emitting an electron and an electron antineutrino in the process. The emitted neutrino carries away a portion of the energy released in the decay, allowing the process to conserve energy, momentum, and angular momentum.

Once a neutrino is emitted, it typically travels vast distances without interacting with matter, due to its weak interaction with other particles. Neutrinos can pass through entire planets or stars almost unimpeded, a fact that has made them extremely difficult to detect. In traditional physics, this weak interaction is attributed to the fact that neutrinos do not have electric charge and interact only through the weak nuclear force, which is much weaker than the electromagnetic and strong nuclear forces. However, in the context of the Gestalt Aether Theory, the emitted neutrino is reinterpreted not as a discrete particle, but as the result of the alignment of the virtual photon aether into a line of force. This line of force is formed when the nucleus is destroyed, and an energetic virtual particle is emitted. The alignment occurs over an incredibly short duration—around 10^{-18} seconds—during which the line of force exists before dissipating. This line of force, according to the Gestalt Aether View, carries away a portion of the energy released during the nuclear interaction. Its short existence and minimal interaction with surrounding matter explain why neutrinos are usually harmless as they pass through normal matter, including our bodies. The theory suggests that the energy associated with this line of force is so fleeting and weak that it has negligible effects on the atoms and molecules it passes through.

Due to the fact that lines of force resulting from neutrino release, are electrically neutral and massless, they have little to no interaction with matter, passing right through atoms without interaction. Despite their weak interaction, neutrinos can occasionally interact with matter under the right circumstances. In rare cases, when a neutrino's line of force passes extremely close to an atomic nucleus, the alignment of the virtual photon aether can become significant enough to disrupt the nucleus. This disruption can lead to the destruction of the nucleus, causing it to emit gamma radiation—a high-energy form of electromagnetic radiation. The emission of gamma rays is a detectable event, and it is through such interactions that neutrinos are sometimes observed.

In traditional physics, such interactions are rare because of the neutrino's weak interaction cross-section, meaning that the probability of a neutrino interacting with a nucleus is extremely low. However, when it does occur, the energy transferred from the neutrino to the nucleus can lead to observable effects, such as the emission of gamma rays or the transformation of a neutron into a proton (or vice versa) in the nucleus.

In environments with extreme densities, such as neutron stars, the density of neutrinos—and thus the density of these lines of force—can become extraordinarily high. In these settings, the cumulative effect of numerous neutrino interactions might contribute to a form of gravity different from ordinary gravity, as suggested by the Gestalt Aether View. This enhanced gravitational effect could be strong enough to lead to the formation of black holes, where the aligned virtual photon aether creates an intense gravitational well from which nothing, not even light, can escape. This theory also implies that the virtual photon aether is not only responsible for the weak interaction we observe with neutrinos but also plays a critical role in the extreme gravitational forces observed in such dense objects. It posits that the alignment of the virtual photon aether by neutrinos might amplify gravity in regions with high neutrino production, potentially contributing to the gravitational collapse into a black hole.

Conclusion

In summary, neutrinos, according to both standard physics and the Gestalt Aether View, are born out of nuclear reactions involving the weak nuclear force, where they carry away a fraction of the energy released in these processes. The Gestalt Aether View, however, reinterprets neutrinos as lines of force formed by the alignment of the virtual photon aether, existing briefly and typically having minimal effects on matter. This line of force theory also suggests that under certain conditions, such as in the vicinity of dense nuclei or in extreme environments like neutron stars, these neutrino lines of force can interact with matter, leading to detectable effects like gamma radiation and possibly influencing gravity in ways not explained by standard models. This alternative perspective offers an intriguing reinterpretation of neutrinos, challenging us to consider the deeper mechanisms that might underlie their mysterious behavior.

Interaction and Detection of Neutrino Lines of Force

In the Gestalt Aether View, when a neutrino is "emitted" during a nuclear reaction, the event results in the creation of a line of force within the virtual photon aether. This line of force is a one-off interaction, distinct from the continuous flow of energy seen in electromagnetic radiation. In electromagnetic interactions, hundreds of trillions of photons travel along a line of aligned virtual photons, transferring energy across space. However, in the case of a neutrino, the entire energy of the neutrino is concentrated in a single event along a line of force within the virtual photon aether. This concentration of energy within a single line of force implies that the energy is significant but fleeting, as the alignment exists only for a brief moment. As this line of force extends through space, it typically encounters no significant interaction with matter. The energy is carried along the aligned virtual photons, but in the vast majority of cases, it passes through planets, stars, and other celestial bodies without causing any effect. The virtual photons realign quickly after the neutrino's energy has been transmitted, making the line of force transient. The reason nothing happens in most cases is that the conditions required for an interaction are extraordinarily rare. For an interaction to occur, the line of force must pass extremely close to a nucleus, aligning in just the right way to transfer sufficient energy to disrupt the nucleus. In most scenarios, the line of force simply travels through matter without incident, because the energy, while concentrated, is not enough to overcome the forces within the atomic nuclei it passes by.

Implications for Neutrino Detection

This understanding provides a twist on why neutrinos are so elusive. In traditional physics, their weak interaction with matter is attributed to the weak nuclear force and the fact that they have no electric charge. In the Gestalt Aether View, the elusiveness is due to the nature of the one-off interaction—the entire energy of the neutrino is carried in a single, concentrated pulse along a line of aligned virtual photons. Because this line of force is so fleeting and only interacts under very specific conditions, neutrinos rarely cause any detectable effects. However, in those rare instances when a neutrino line of force does interact with a nucleus, the result can be dramatic. The energy transfer can disrupt the nucleus, leading to its destruction and the emission of gamma radiation. This gamma radiation can then be detected, providing evidence of the neutrino's passage.

In summary, the Gestalt Aether View reinterprets the behavior of neutrinos as one-off energy transfers along lines of force created in the virtual photon aether. Unlike electromagnetic radiation, which involves continuous streams of photons, a neutrino's energy is carried in a single, concentrated pulse along this transient line of force. While these lines of force generally pass through matter without interaction, their rare encounters with atomic nuclei can result in detectable events, such as gamma radiation emission. This perspective offers a new understanding of neutrinos' elusive nature and the conditions under which they might be detected.

Chapter 8 : Magnetism and super conductivity.

The Nature and Origins of Magnetism: A Critical Examination

The act of coming face to face with a magnet is almost a rite of passage for most boys. This is an object that possesses unknown magical properties. That can attract other objects, made of iron, without any apparent motivational force being present. As one grows older, one notices the similarities behind a current carrying wire and a permanent magnet. After the differences are explained one tends to lose interest, putting the similar property of a wire carrying an electric current and a permanent magnet aside and carrying on with other things.

Magnetism has been recognized since ancient times, evidenced by the discovery of magnetite, a naturally occurring iron oxide with magnetic properties. Similarly, the phenomenon of static electricity was also known in antiquity, demonstrated by the fact that amber, when rubbed against fur, could attract small objects. The similarities between static electricity and magnetism, such as the attraction of like poles and the repulsion of opposite poles, were noted by early observers.

However, it was not until the mid-nineteenth century that the connection between electricity and magnetism became apparent. James Clerk Maxwell ultimately unified these two forces into the concept of the electromagnetic field, marking a significant milestone in the understanding of physics. Today, it is widely accepted that every form of magnetism—whether in the magnetic field around an electromagnet, the magnetism in a permanent magnet, or in a piece of magnetite—is caused by the passage of an electric current passing through these substances. Despite this understanding, the fundamental causes of magnetism remain somewhat elusive.

Magnetism primarily manifests in metals, and modern physics offers three main explanations for its existence. First, electrons, which are charged particles, orbit the nucleus of an atom, generating magnetic fields as they move. In some metals, these fields align, resulting in magnetism, while in others, they cancel out, leading to non-magnetic properties. Second, the electron's spin as it orbits the nucleus is also believed to contribute to magnetism. Third, the spin of the nucleus itself might play a role in generating magnetic fields.

While these explanations might seem logical, they are not entirely satisfactory. For instance, the wave-particle duality of electrons was introduced to explain why electrons do not radiate energy as they orbit the nucleus. Yet, this concept is often conveniently resurrected to explain magnetic fields without fully considering its broader implications. This raises questions about the consistency of such explanations, especially in light of Max Born's observation in **The Restless Universe (1956)**, where he states, "One should not imagine that there is anything in the nature of matter actually rotating," referring to the in-aptly named 'spin' of electrons.

According to Gestalt Aether Theory, the electromagnetic fields around a wire carrying a current are produced by free electrons within the conductor emitting and then immediately reabsorbing 'conduction photons' with a wavelength of 1.2×10^{-6} meters. These photons exit the conductor only to re-enter it, being absorbed by the same or another electron. This process aligns with Heisenberg's Uncertainty Principle, which suggests that if an action occurs rapidly enough—in this case, in approximately 10^{-15} seconds—the laws of conservation of momentum and energy can be bypassed. The virtual photons of the virtual photon field, or aether, then align in the direction of the real photon's propagation as it exits and re-enters the conductor, creating the characteristic lines of force around a current-carrying wire. Thus, according to Gestalt Aether Theory, magnetism is merely a manifestation of the 'virtual photon field' or aether.

This notion raises further questions: how does this theory account for the phenomenon of permanent magnets? The Gestalt Aether theory perspective provides a compelling explanation for the electromagnetic fields around conductors, but it does not fully address the complexities of permanent magnetism, leaving room for further exploration and understanding.

Magnetism mainly manifests itself in metals. Modern physics reasons that because it has been established that a moving or accelerating charge gives rise to an electromagnetic field, there must be three causative factors for the existence of magnetism:

It is known that electrons are charged particles that orbit around the nucleus; thus, it follows that all atoms generate magnetic fields. In some metals, the fields generated in this manner line up, making these metals magnetic. In others, they cancel each other out, and these metals do not exhibit magnetism.

1. The electron, as it orbits around the nucleus, spins on its axis, and this spin gives rise to magnetism.
2. The nucleus itself is spinning, and this might also give rise to magnetism.

Yet although the above explanations of how magnetic fields are generated in both permanent magnets and temporary electromagnets might seem to make sense, they are vaguely unsatisfactory. For instance, the whole concept of wave-particle duality came about as an explanation of why electrons do not radiate as they orbit the nucleus. Does it make sense, then, to suddenly resurrect this explanation of magnetic fields when it is needed, conveniently ignoring the implications of what this might mean in terms of related phenomena? This is especially so when one considers Max Born's statement when referring to the 'spin' of electrons: "One should not imagine that there is anything in the nature of matter actually rotating……" (Max Born, The Restless Universe, 1956).

The Gestalt Aether Theory of Magnetism

Gestalt Aether Theory holds that magnetic (electromagnetic) fields around a wire carrying a current are caused by free electrons within the conductor emitting and immediately reabsorbing 'conduction photons' having a wavelength of 1.2×10^{-6} meters. These photons exit the conductor and then immediately re-enter it to be absorbed by the same or another electron, in keeping with Heisenberg's Uncertainty Principle, which states that if an action takes place fast enough—in this case, in around 10^{-15} seconds—then the laws of conservation of momentum and energy can be bypassed. The virtual photons of the virtual photon field (aether) then line up in the direction of the propagation of the real photon as it exits and enters the conductor, giving rise to the typical lines of force evident around a conductor carrying a current. Thus, according to Gestalt Aether Theory, the phenomenon of magnetism is merely the manifestation of the 'virtual photon field' or aether. How does this account for the phenomenon of permanent magnets?

In order to understand the Gestalt Aether Theory explanation for the phenomenon of permanent magnetism, it is necessary to understand a little about the structure of metals, in particular the crystalline lattice structure of metals. The atomic arrangement within a crystal is called crystal structure, which equates with a periodic arrangement of points in space about which these atoms are located. This is known as a space lattice. A space lattice is defined as an infinite array of points in three dimensions in which every point has surroundings identical to every other point in the array. A space lattice can be defined by referring to a unit cell. The unit cell is the smallest unit that, when repeated in space indefinitely, generates the space lattice.

There are only 14 distinguishable ways of generating points in three-dimensional space. These 14 space lattices are known as Bravais lattices, and they belong to seven crystal systems. The seven crystal structures are:

- Cubic
- Tetragonal

- Orthorhombic
- Monoclinic
- Triclinic
- Trigonal
- Hexagonal

The use of X-rays for the investigation of crystals has led to a great understanding of the structure of these crystalline structures since the X-ray wavelength is similar to the size of the crystalline sides. It has been found that most metals prefer a simple cubic structure since it is the strongest and most stable structure. Here things become very interesting. For instance, it is found that the property of electrical conduction is almost wholly dependent on the kind of crystalline lattice structure found in the metal. For instance, all of the three top conductors in metal, namely silver, copper, and aluminum, have the same crystalline structure, namely a face-centered cubic (FCC) structure.

Face Centred Cubic structure

In the Face-Centered Cubic structure, all positions in the cube lattice are packed with atoms. This means that there is an abundance of electrons present, and the valence electrons are only loosely attached, enabling the presence of many free electrons within the conductor. This gives such metals very large bandwidth, making them very good conductors of electricity. Iron, which has a Body-Centered Cubic (BCC) structure, has only about one-fortieth the conductivity of copper. It is also very susceptible to erosion by oxidation.

Body Centred Cubic Structure

The purity of the crystalline structure of the metal contributes greatly to the conductivity of the metal. For instance, steel, which is an alloy, containing different amounts of manganese, molybdenum, chromium, and nickel, as well as carbon, has a lattice structure that is not uniform, leading to poor conductivity.

Structure of steel crystal

A study of the properties of the crystalline structures of metals also reveals that metals susceptible to permanent magnetism possess similar crystalline structures to each other. For instance, the ability of a metal to have more than one structure in its solid state is called polymorphism. If the change in structure is reversible, it is known as allotropy. Iron, cobalt, and nickel are all ferromagnetic and possess the same crystalline structure, namely the Body-Centered Cubic structure. All of these metals also demonstrate the property of allotropy. Thus, it is possible to gain some idea of how these metals become magnetized. Several criteria must be fulfilled. Permanent magnetism is only possible using direct current; the use of alternating current does not result in the formation of permanent magnets. Furthermore, this current must flow continuously for a certain period of time for magnetization to occur. The substance to be magnetized must be either iron, nickel, or cobalt, or an alloy containing one or more of these elements, such as aluminum, nickel, and cobalt to make Alnico permanent magnets. Here is what happens during the process of creating a permanent magnet, according to Gestalt Aether Theory. Once a difference of potential has been established across a conductor, free electrons within the conductor start to emit and absorb photons, resulting in lines of force forming around the conductor, with each line of force containing the energy of a single 'conduction photon'. It is important to mention that the field around a wire carrying a DC current contains only near-field elements, apart from extremely residual far fields. Far fields exist primarily around conductors carrying alternating current. To make a permanent magnet, it is usual to use a solenoid coil, which gives rise to the magnetic field found around a bar magnet. Depending on the kind of field used in the magnetization process, it is possible to form permanent magnets having differently configured magnetic fields.

In metals that are good conductors and possess a Face-Centered Cubic structure, it has been noted that there are a great many free electrons. While these metals (copper, silver, aluminium) make good conductors, they are paramagnetic or have extremely weak magnetic properties. In these metals, electricity is conducted through the emission and absorption of 'conduction photons'. However, since the photons are emitted and absorbed in an extremely chaotic manner, with one electron emitting a photon that another electron, which has also emitted a photon, absorbs and so on, it is not possible for the conduction of electricity to impose any lasting order on the emission and absorption process. In metals that are based on the Body-Centered Cubic structure, on the other hand, there are comparatively few free electrons, and those that do exist are more or less bound to the atom. Hence, in those metals that possess a Body-Centered Cubic crystalline structure and exhibit the property of allotropy, it is thought that the conduction of electricity takes place by emission and absorption of 'conduction photons' by bound electrons in the outer orbits of atoms that would normally randomly substitute valence electrons. When a current is established, these outer orbit electrons are permanently assigned to emitting and absorbing 'conduction photons', resulting in the establishment of a flow of current. As the process continues, the metals undergo polymorphism wherein the structure of the crystalline lattice of the metal itself undergoes changes to produce the most efficient configuration for the emission and absorption of photons. Once the metal undergoes polymorphism, changing into the configuration most suited to sustain the emission and absorption of 'conduction photons'. The absorption and emission process takes place in repetitive cycles, with the electron absorbing and emitting the same 'conduction photon' for long periods of time. When the current is disconnected, the photons continue to be absorbed and emitted in the same pattern, giving rise to the distinctive lines of force that are found around permanent magnets. This is how, permanent magnets are formed.

Hence, Gestalt Aether Theory states that permanent magnetism is identical in every way to electromagnetism; there is no difference! Magnetism, is like a form of frozen current, in the connotation of being still not cold. The only difference is that in a permanent magnet, the lines of force (i.e., current-carrying lines of aligned virtual photons) are fixed in permanent loops and thus cannot convey current, although they can induce a current. For instance, when an electrically conductive rod is brought into the vicinity, the lines of force will enter the rod, giving rise to a current. This happens because the atoms in the moving rod are disturbed through the movement of the rod and are therefore receptive to 'conduction photons'. Permanent magnetism is exhibited by metals having a Body-Centered Cubic crystalline lattice structure and which also possess the property of allotropy. Because of this, permanent magnets that are hit sharply or subjected to heat or electricity will lose their magnetism. It is also true that it is impossible to drill through a permanent magnet!

Superconductivity

The topic of superconductivity, while only briefly mentioned here, demonstrates the impressive ability of Gestalt Aether Theory physics to explain complex phenomena. Discovered in 1911 by Dutch physicist Heike Kamerlingh Onnes and his team—Cornelis Dorsman, Gerrit Jan Flim, and Gilles Holst—superconductivity initially sparked hopes for lossless electrical transmission. However, there was a significant challenge: superconductors require extremely low temperatures to function. Conventional superconductors, for instance, need to be cooled to around 39 kelvins (approximately -234°C or -389°F). The mercury wire used by Kamerlingh Onnes functioned at even lower temperatures, below 4.2 K (-269.0°C or -452.1°F). Even so-called "high-temperature" superconductors work below 130 K (-143°C or -225.7°F).

Additionally, superconductors can lose their resistance-free state if exposed to strong magnetic fields or excessive electric current.

Modern superconductors, such as niobium-titanium (NbTi), have improved magnetic load-bearing capabilities, making them useful in applications like maglev trains, proton accelerators (e.g., Fermilab), and MRI machines.

One of the unique properties of superconductors is that when an electric current flows through a superconducting wire, it encounters no resistance and loses no energy. If this wire is bent into a loop, it can maintain an electrical current indefinitely. Furthermore, if a magnet is placed near a superconductor, it will levitate above it, remaining in this state as long as the superconductor stays cold and powered. Physicists have long been intrigued by these phenomena and have developed complex theories to explain them. According to Gestalt Aether Theory physics, the fundamental charge carriers in superconductors are not electrons but photons. This theory suggests that at the low temperatures of 4.2 K, free electrons in the superconductor are bound as valence electrons. When a current is introduced, these valence electrons continuously absorb and emit photons, forming closed loops within the conductor. As a result, no photons leave the conductor, and no external lines of force are created.

Within a superconductor cooled to its critical temperature, an introduced charge leads to an orderly system where each photon is associated with a specific bound electron, emitting and absorbing conduction photons in a circular pattern within the conductor. Unlike in other materials, no conduction photons exit the superconductor. This unique situation results in a current that circulates endlessly within the wire, absorbed and emitted repeatedly by the same electrons. Consequently, no external lines of force are produced, and external magnetic fields cannot interact with the superconductor. Instead, the superconductor interacts with the virtual photon aether surrounding it, forming an annular pattern of promoted virtual photons above its surface. This interaction explains why an ordinary magnet brought near this field will levitate above the superconductor.

Annular lines of force formed on surface of superconductor

The Gestalt Aether Theory theory of superconductivity highlights two critical points: (a) it validates the idea that photons, rather than electrons, serve as the charge carriers, and (b) it demonstrates that lines of force result from the alignment of virtual photons in the aether, as no such lines appear outside the superconductor once a current is established. The absence of external lines of force is explained by the fact that no photons escape the conductor.

This evidence supports the Gestalt Aether Theory, showing that simple reasoning based on hypotheses in the Newtonian tradition can lead to predictions confirmed by experimental results. In contrast, explanations like Cooper pairing fail to account for these phenomena. The annular pattern of lines of force around the superconductor prevents interaction with the lines of force from an ordinary magnet, creating the appearance that the magnetic lines of force are being repelled by the superconductor. However, the interaction between the permanent magnet and the superconductor is sufficiently strong to cause the magnet to levitate above it and even to turn the whole upside down and reverse the effect with the magnet apparently levitating the superconductor.

Chapter 9 : Conclusion: Toward a New Paradigm in Physics

Newtonian concepts of absolute space and time:

Newton's ideas of absolute space and time form a cornerstone of classical physics, providing a framework that shaped scientific thought for centuries. These concepts, though seemingly intuitive, were groundbreaking in their time and laid the foundation for what would become the dominant worldview in physics until the early 20th century. Isaac Newton, in his Philosophiæ Naturalis Principia Mathematica (1687), introduced the concepts of absolute space and absolute time as key elements of his laws of motion and universal gravitation. According to Newton, absolute space is a fixed, immovable stage upon which the events of the universe unfold. It exists independently of any objects or events within it—a vast, unchanging arena that provides the backdrop for all physical phenomena. Similarly, absolute time, in Newton's view, flows uniformly and inexorably, independent of the events that occur within it. This conception of time as an ever-present, continuous stream was crucial to his mathematical descriptions of motion and the laws governing it.

Newton's ideas were revolutionary because they diverged significantly from earlier conceptions of space and time. Prior to Newton, many philosophers, including Aristotle, saw space as a relational concept—defined only by the positions and distances between objects. Time, likewise, was often seen as a measure of change, inseparable from the events it marked. Newton, however, posited that space and time existed as entities in their own right, not merely as relationships between objects or events. These ideas were not merely abstract philosophical musings; they were deeply intertwined with Newton's physical theories. The notion of absolute space allowed Newton to formulate his laws of motion in a way that was consistent across the universe. For instance, his first law—the law of inertia—asserts that an object in motion remains in motion with a constant velocity unless acted upon by an external force. This law implicitly relies on the existence of a fixed space against which motion can be measured. Without absolute space, the very concept of velocity would be relative, and Newton's laws would lose their universal applicability. Similarly, absolute time was essential for Newton's formulation of the laws of motion and gravitation. The concept of a uniform, unchanging time allowed him to describe the motions of planets and objects on Earth with the same set of equations. Newton's law of universal gravitation, which posits that every mass in the universe attracts every other mass with a force proportional to their masses and inversely proportional to the square of the distance between them, requires a consistent measure of time to predict the motions of celestial bodies accurately.

Historically, Newton's ideas of absolute space and time were both a response to and a departure from the work of earlier scientists like Galileo Galilei and René Descartes. Galileo's principle of inertia, which stated that an object in motion stays in motion unless acted upon by an external force, hinted at the idea of a uniform motion but did not fully account for the nature of space and time. Descartes, on the other hand, had proposed a mechanistic view of the universe where space was filled with matter, and motion was the result of interactions between these material bodies. Newton's absolute space and time provided a more abstract, yet mathematically precise, foundation for understanding the physical world. Newton's ideas were not without controversy. The concept of absolute space, in particular, was criticized by some of his contemporaries and later philosophers. The German philosopher and mathematician Gottfried Wilhelm Leibniz, for example, argued against the notion of absolute space, asserting that space was a relational concept, defined by the relationships between objects rather than an independent entity. Leibniz believed that the idea of an empty space, independent of the objects within it, was meaningless.

Despite these criticisms, Newton's concepts of absolute space and time became deeply entrenched in the scientific worldview of the 18th and 19th centuries. They were integral to the development of classical mechanics and provided the framework for understanding the physical universe. It was not until the advent of Einstein's theory of relativity in the early 20th century that Newton's notions of absolute space and time were fundamentally challenged. Einstein's evolved a theory where space and time are not absolute but are intertwined and relative to the observer, marking a profound shift in our understanding of the universe. In conclusion, Newton's ideas of absolute space and time were postulates of classical physics, forming the bedrock upon which his laws of motion and universal gravitation were built. These concepts allowed Newton to describe the physical universe with unprecedented precision and laid the foundation for centuries of scientific progress. Although later developments in physics would reveal the limitations of

Newton's framework, his contributions remain a monumental achievement in the history of science.

Dark matter shares several characteristics that were once attributed to the classical concept of the aether, making it a compelling candidate for reconsidering the role of a medium-like substance in the universe. Dark matter is a mysterious substance that does not emit, absorb, or reflect light, making it invisible to current detection methods. However, its presence is inferred from its gravitational effects on visible matter, such as galaxies and clusters of galaxies. Dark matter appears to permeate the universe, forming a kind of scaffolding that dictates the large-scale structure of cosmic matter. This pervasive and invisible substance, which interacts with the rest of the universe primarily through gravity, exhibits the same elusive properties that were once attributed to the aether.

Historically, the aether was conceived to be an all-pervading medium through which light waves were thought to propagate, similar to how sound waves travel through air. This concept was ultimately discarded after experiments like the Michelson-Morley experiment failed to detect its presence. However, dark matter reintroduces the idea of an invisible, omnipresent substance that influences the behavior of matter and energy on a cosmic scale.

If dark matter is the modern equivalent of the aether, it suggests that space might not be as empty or as abstract as it appears in the relativistic framework. Instead, space could be filled with a complex and subtle medium that affects the motion of galaxies, the propagation of gravitational waves, and conceivably the propagation and behavior of light and other electromagnetic radiation under certain conditions. This could imply that space and time, far from being purely relative as Einstein proposed, have a more absolute character rooted in the properties of dark matter. The concept of dark matter as a new form of aether challenges the current understanding of the universe and invites a reconsideration of the fundamental nature of space and time. If dark matter is indeed the medium that fills space, it could provide a preferred frame of reference—something that relativity explicitly denies.

In light of this, it seems plausible to argue that the existence of dark matter, with its vast influence on the cosmos, is seen as a modern vindication of the aether concept, albeit in a form that is vastly different from the 19th-century understanding of a luminiferous aether. This view aligns with the idea that Newton's absolute space and time might not be entirely obsolete but rather misunderstood or incomplete in the absence of an understanding of dark matter. Therefore, rather than dismissing the possibility of a medium like the aether, it might be more appropriate to consider dark matter as a potential candidate that fulfills a similar role in a contemporary context. This perspective opens up new avenues for exploring the fundamental nature of the universe and could lead to a deeper understanding of both dark matter and the fabric of space and time. The search for a complete theory that reconciles these ideas continues, and dark matter may well hold the key to resolving the century-old debate between Newtonian absolutes and Einsteinian relativity.

Frames of reference:

The idea that Dark matter serves as a medium for light and all kinds of electromagnetic radiation, in the same way the aether was supposed to, is validated to an extent by the fact that dark matter allows the free passage of all types of electromagnetic radiation, in all directions, without offering any resistance or interference. From the Gestalt Aether theory point of view its role in the gravitational dynamics of the universe hints at an underlying structure that is seen as a modern parallel to the aether.

Frames of reference: A café in New York:

Imagine that it is a Thursday and I am going to meet someone at a cafe in New York, exactly one week later I again go to the same cafe to visit the same person, it should not be assumed that I have returned to the same place because during that time, the earth has been rotating on its axis at a speed of about 1000 kmph and also travelling through space at 60,000 kmph and the cafe is now 2.5 million kilometres distant from where it was last Thursday. The important thing to understand is that all this movement can be considered to be superficial and irrelevant because, during that time the earth has also been travelling through the aether with zero interaction. During that time the aether has remained absolutely still and stationary. The same holds good for time, because it is the aether which serving as the medium for light that is responsible for causality. The aether therefore is also responsible for the arrow of time. This analogy illustrates the concept of absolute space and time in relation to the aether, complementing Newton and presenting a compelling argument that challenges the relativistic framework established by Einstein.

Imagine standing in a café in New York on a Thursday, and exactly one week later, returning to the same café to meet the same person. In a purely Newtonian sense, it seems as though you've returned to the same place at the same time. However, from a relativistic perspective, the café is no longer in the same position in the universe. During that week, the Earth has traveled a significant distance through space—around 2.5 million kilometers— due to its rotation on its axis, its orbit around the Sun, the Solar System's movement through the galaxy, and the galaxy's journey through the universe. Despite this vast movement, the café appears unchanged, and the meeting happens as planned. This circumstance strongly suggests that there is something constant, something absolute, which allows for the seamless continuity of events despite the relative motion of everything in the universe.

In the scenario described, this "something" could be the aether—a stationary, absolute frame of reference that remains unchanged regardless of the Earth's movement through space. The aether, in this view, would serve as a universal backdrop against which all motion occurs, providing a fixed stage for the unfolding of events. Despite the Earth's constant motion, the aether remains unaffected, ensuring that the café, though physically displaced in the universe, remains consistent in relation to this absolute space. The implication of such a concept is profound. If the aether exists as a stationary medium through which the Earth and all objects move, it would mean that space and time are not relative, as Einstein proposed, but absolute, as Newton believed. In this framework, the aether provides an unchanging reference point, ensuring that events in the universe occur in a consistent manner, irrespective of the relative motion of objects. This would mean that the café, despite its movement through space, remains "in the same place" within the absolute space defined by the aether.

Furthermore, the point about the aether being responsible for causality and the arrow of time is also validated. Taking the aether to be the medium through which light propagates, automatically implies it would also govern and limit the speed of light, which is central to the concept of causality. The constancy of the speed of light ensures that cause and effect relationships remain consistent across the universe, defining the flow of time from past to future. In this sense, the aether could be seen as not just the fabric of space, but also as the medium that maintains the directionality of time, giving rise to the "arrow of time" that we experience.

This idea contrasts sharply with the relativistic notion that time is relative and can vary depending on the observer's frame of reference. Instead, it suggests that time, like space, is absolute and governed by the properties of the aether. The aether would then not only dictate the propagation of light but also the uniform flow of time, ensuring that events unfold in a consistent, ordered manner across the universe. In essence, this view suggests that the aether, as a stationary and absolute medium, could be the key to understanding the true nature of space and time. It challenges the relativistic view by proposing that space and time are constants, rooted in the existence of an aether that remains unaffected by the motion of objects within it. This perspective offers a return to a more Newtonian understanding of the universe, where absolute space and time are fundamental properties, anchored by the presence of the aether.

Gestalt Aether theory might play a crucial role in the reconciliation between classical and relativistic physics.

As we reach the conclusion of our exploration into the theory that has come to be known as Gestalt Aether, or

Neo-Classical Physics, we stand at the threshold of a profound transformation in our understanding of the universe. This journey has not been about merely challenging the established paradigms of quantum mechanics and relativity but rather about offering a cohesive and intuitive framework that seeks to harmonize the seemingly disparate phenomena observed in nature. The central tenet of this theory is the existence of a pervasive medium, the so-called virtual photon aether or dark matter, that fills the cosmos and provides the stage upon which all physical interactions take place. Unlike the classical aether, dismissed by Einstein over a century ago, this new aether is not a substance in the traditional sense but a field of virtual photons that permeates every corner of the universe. It is this field that facilitates the propagation of electromagnetic radiation, governs gravitational interactions, and underpins the very laws of thermodynamics.

Through the lens of this theory, we have revisited the nature of the electron, reimagined as a solid particle that regulates its energy through the emission and absorption of virtual photons. We have examined the propagation of light, not as a wave traveling through empty space but as a process involving the sequential promotion of virtual photons to real photon status, creating a wave-front that adheres to the inverse square law. Gravity, too, has been reinterpreted as an emergent property of the alignment of these virtual photons, a force that manifests through the shortest possible distance between objects. One of the most compelling aspects of this theory is its ability to provide a unified explanation for phenomena that have long been treated as separate domains of physics. By positioning the virtual photon aether as the medium through which both electromagnetic and gravitational forces operate, we can draw connections between these fundamental forces in ways that previous theories could not. This unification does not rely on abstract mathematical constructs or higher dimensions but on the physical processes occurring within the aether itself.

Moreover, this theory brings us back to a more classical understanding of the universe, one in which the principles of energy conservation and the inverse square law are not just mathematical conveniences but deeply rooted in the fabric of reality. It offers a return to a form of physics that is grounded in observable phenomena and intuitive reasoning, a physics that does not shy away from the complexities of the universe but seeks to explain them in a coherent and comprehensible manner. Yet, this is not a return to the past but a reimagining of the future. The Gestalt Aether Theory, or Neo-Classical Physics, does not seek to discard the advancements of the last century but to build upon them, integrating the insights of quantum mechanics and relativity into a broader, more inclusive framework. It acknowledges the successes of these theories while addressing their shortcomings, particularly their reliance on non-physical dimensions and probabilistic interpretations that many find unsettling.

As an interesting corollary to this theory, it must be noted that beyond providing novel and effectual explanations for the flow of current along wires, the formation and propagation of radio waves, the propagation of light, and most importantly, an explanation of gravity, this theory also offers groundbreaking insights into the phenomena of magnetism and superconductivity—two areas that have long perplexed scientists. Take the case of magnetism, for instance. Traditionally in the pre-quantum mechanics days, magnetism has been attributed to the electron's orbit around the nucleus, and later to the electron's intrinsic spin. These explanations, however, have always been rather unsatisfactory. The notion that the delicate balance of forces and the stable configurations of electrons within atoms could somehow produce a pervasive and powerful force like magnetism appears, on closer examination, to be a rather tenuous proposition. These conventional explanations lack the clarity and coherence that one expects from a truly robust scientific theory. Indeed, magnetism remains one of the great mysteries of physics, a phenomenon that has never been fully solved or adequately explained within the framework of traditional theories.

Gestalt Aether Theory, claims to do exactly that. By rethinking the role of virtual photons and the interaction between particles at a fundamental level, it offers a fresh perspective that not only makes sense but also aligns with observable phenomena in a way that traditional theories have struggled to achieve. Magnetism, in this view, is not merely a byproduct of electron motion or spin but is instead a natural consequence of the interactions within the virtual photon aether—a concept that bridges the gap between classical physics and quantum mechanics. In a similar vein, the explanation of superconductivity through the formation of Cooper pairs—where electrons supposedly form pairs to move through a lattice without resistance—has always seemed counterintuitive when viewed from the standpoint of classical physics. This explanation, while mathematically sound, conflicts with the fundamental principles that govern the behavior of charged particles. The notion that two negatively charged electrons would pair up, rather than repel each other, defies the very laws of electrostatics that are otherwise inviolable. Yet, this is the accepted explanation within the current paradigm.

Once again, Gestalt Aether theory offers a solution that overcomes these objections. By revisiting the interactions at play on the quantum level and incorporating the concept of the virtual photon aether, it provides a more cohesive and physically intuitive explanation for superconductivity. This approach not only aligns with classical principles but also offers a more comprehensive understanding of how and why superconductivity occurs, especially at low temperatures. In summary, this theory

302

does not merely tweak existing ideas but rather revolutionizes our understanding of fundamental forces and interactions. It provides new, coherent explanations for phenomena that have long been considered enigmatic, such as magnetism and superconductivity, and does so in a way that is consistent with both classical and quantum principles. The detailed explanation of how this theory addresses magnetism and superconductivity will be explored elsewhere, as this book has already explained enough of its foundational ideas.

Today we find ourselves on the cusp of a new era in physics, one that bridges the gap between the microscopic and macroscopic worlds, between the tangible and the intangible, and between the known and the unknown. In the end, the name we give to this theory is less important than the ideas it represents. Whether we call it Gestalt Aether, Neo-Classical Physics, or something else entirely, what matters is the pursuit of a deeper understanding of the universe. And in that pursuit, we may yet find the answers to some of the most fundamental questions of existence. The journey has just begun.

Index

306

310

Made in United States
North Haven, CT
22 October 2024

59321553R00183